NATURE

GUIDE *to*

ATLANTIC

CANADA

Erin McCloskey & Gregory Kennedy

LONE
PINE

Lone Pine Publishing

Lone Pine Publishing
2311–96 Street
Edmonton, Alberta T6N 1G3

Website: www.lonepinepublishing.com

Library and Archives Canada Cataloguing in Publication

McCloskey, Erin, 1970–

   Nature guide to Atlantic Canada / Erin McCloskey and Gregory Kennedy.

Includes bibliographical references and index.

ISBN 978-1-55105-612-8

   1. Natural history—Atlantic Provinces—Guidebooks. I. Kennedy, Gregory, 1956– II. Title.

QH106.2.A84M34 2012   577.09715   C2011-907597-0

Editorial Director: Nancy Foulds
Project Editor: Nicholle Carrière
Editorial: Nicholle Carrière, Sheila Quinlan, Wendy Pirk
Production Coordinator: Gene Longson
Layout & Production: Volker Bodegom, Janina Kürschner
Cover Design: Gerry Dotto

Cover Images: Frank Burman, Ted Nordhagen, George Penetrante, Gary Ross, Ian Sheldon

Illustrations: A complete list of illustration credits appears on page 4, which constitutes an extension of this copyright page.

Disclaimer: This guide is not intended to be a "how to" reference guide for food or medicinal uses of plants. We do not recommend experimentation by readers, and we caution that a number of plants in Atlantic Canada, including some used traditionally as medicines, are poisonous and harmful.

We acknowledge the financial support of the Government of Canada through the Canada Book Fund (CBF) for our publishing activities.

 Canadian   Patrimoine
Heritage   canadien

PC: 16

TABLE OF CONTENTS

ILLUSTRATION CREDITS

ACKNOWLEDGEMENTS

For their time in helping select the species represented in this guide, thanks go to Melanie Priesnitz, conservation horticulturist and PlantWatch coordinator at the KC Irving Environmental Science Centre and the Harriet Irving Botanical Gardens; Andrew Hebda, zoology curator at the Nova Scotia Museum; and Dr. Donna Giberson at the University of Prince Edward Island.

Also, thank you to Nicholle Carrière for being such a talented and helpful editor.

–Erin McCloskey

moose

MAMMALS

Blue Whale
p. 49

Fin Whale
p. 49

Humpback Whale
p. 50

Atlantic Minke Whale
p. 50

Sei Whale
p. 51

North Atlantic Right Whale
p. 51

Long-finned Pilot Whale
p. 52

Beluga
p. 52

Orca
p. 53

Risso's Dolphin
p. 53

Bottlenose Dolphin
p. 54

Short-beaked Common Dolphin
p. 54

Atlantic White-sided Dolphin
p. 55

Harbour Porpoise
p. 55

Harbour Seal
p. 56

Ringed Seal
p. 56

Grey Seal
p. 57

Harp Seal
p. 57

Bearded Seal
p. 58

Hooded Seal
p. 58

Moose
p. 59

White-tailed Deer
p. 59

Woodland Caribou
p. 60

Mountain Lion
p. 60

Lynx
p. 61

Bobcat
p. 61

Grey Wolf
p. 62

Coyote
p. 62

Red Fox
p. 63

Arctic Fox
p. 63

Black Bear
p. 64

Polar Bear
p. 64

Wolverine
p. 65

Marten
p. 65

Fisher
p. 65

Long-tailed weasel
p. 66

MAMMALS

Short-tailed weasel
p. 66

Mink
p. 66

Northern River Otter
p. 67

Striped Skunk
p. 67

Raccoon
p. 67

Arctic Hare
p. 68

Snowshoe Hare
p. 68

Beaver
p. 68

Porcupine
p. 69

Woodchuck
p. 69

Red Squirrel
p. 69

Northern Flying Squirrel
p. 70

Eastern Chipmunk
p. 70

Meadow Vole
p. 70

Southern Bog Lemming
p. 71

Common Muskrat
p. 71

Norway Rat
p. 71

Deer Mouse
p. 72

House Mouse
p. 72

Woodland Jumping Mouse
p. 72

Star-nosed Mole
p. 73

Masked Shrew
p. 73

Big Brown Bat
p. 73

Eastern Pipistrelle
p. 74

MAMMALS

Eastern Red Bat
p. 74

Hoary Bat
p. 74

Little Brown Bat
p. 75

Silver-haired Bat
p. 75

BIRDS

Canada Goose
p. 78

Wood Duck
p. 78

American Black Duck
p. 78

Green-winged Teal
p. 79

Common Eider
p. 79

Common Merganser
p. 79

Ring-necked Pheasant
p. 80

Ruffed Grouse
p. 80

Willow Ptarmigan
p. 80

Common Loon
p. 81

Pied-billed Grebe
p. 81

Northern Gannet
p. 81

Double-crested Cormorant
p. 82

Great Cormorant
p. 82

American Bittern
p. 82

Great Blue Heron
p. 83

Green Heron
p. 83

BIRDS

Black-crowned Night-heron
p. 83

Osprey
p. 84

Bald Eagle
p. 84

Northern Harrier
p. 84

Sharp-shinned Hawk
p. 85

Northern Goshawk
p. 85

Broad-winged Hawk
p. 85

Red-tailed Hawk
p. 86

American Kestrel
p. 86

Merlin
p. 86

Virginia Rail
p. 87

Sora
p. 87

Semipalmated Plover
p. 87

Killdeer
p. 88

Spotted Sandpiper
p. 88

Greater Yellowlegs
p. 88

Ruddy Turnstone
p. 89

Sanderling
p. 89

White-rumped Sandpiper
p. 89

Dunlin
p. 90

Wilson's Snipe
p. 90

Black-headed Gull
p. 90

Herring Gull
p. 91

Great Black-backed Gull
p. 91

Common Tern
p. 91

Razorbill
p. 92

Black Guillemot
p. 92

Atlantic Puffin
p. 92

Rock Pigeon
p. 93

Mourning Dove
p. 93

Black-billed Cuckoo
p. 93

Great Horned Owl
p. 94

Snowy Owl
p. 94

Barred Owl
p. 94

Long-eared Owl
p. 95

Short-eared Owl
p. 95

Northern Saw-whet Owl
p. 95

Common Nighthawk
p. 96

Ruby-throated Hummingbird
p. 96

Belted Kingfisher
p. 96

Downy Woodpecker
p. 97

Black-backed Woodpecker
p. 97

Northern Flicker
p. 97

Pileated Woodpecker
p. 98

Olive-sided Flycatcher
p. 98

Alder Flycatcher
p. 98

Eastern Kingbird
p. 99

Northern Shrike
p. 99

BIRDS

Red-eyed Vireo
p. 99

Grey Jay
p. 100

Blue Jay
p. 100

Common Raven
p. 100

Horned Lark
p. 101

Tree Swallow
p. 101

Barn Swallow
p. 101

Black-capped Chickadee
p. 102

Red-breasted Nuthatch
p. 102

Winter Wren
p. 102

Golden-crowned Kinglet
p. 103

Eastern Bluebird
p. 103

Veery
p. 103

Hermit Thrush
p. 104

Wood Thrush
p. 104

American Robin
p. 104

Gray Catbird
p. 105

Northern Mockingbird
p. 105

Cedar Waxwing
p. 105

Ovenbird
p. 106

Northern Waterthrush
p. 106

Black-and-white Warbler
p. 106

American Redstart
p. 107

Northern Parula
p. 107

Yellow Warbler
p. 107

Savannah Sparrow
p. 108

Song Sparrow
p. 108

White-throated Sparrow
p. 108

Dark-eyed Junco
p. 109

Northern Cardinal
p. 109

Snow Bunting
p. 109

Red-winged Blackbird
p. 110

Common Grackle
p. 110

Brown-headed Cowbird
p. 110

Purple Finch
p. 111

White-winged Crossbill
p. 111

Common Redpoll
p. 111

Pine Siskin
p. 112

American Goldfinch
p. 112

House Sparrow
p. 112

Red-spotted Newt
p. 114

Blue-spotted Salamander
p. 114

Spotted Salamander
p. 114

Northern Two-lined
Salamander p. 115

Four-toed Salamander
p. 115

Red-backed Salamander
p. 115

Spring Salamander
p. 116

AMPHIBIANS & REPTILES

American Toad
p. 116

Northern Spring Peeper
p. 116

Wood Frog
p. 117

Northern Leopard Frog
p. 117

Green Frog
p. 117

Bullfrog
p. 118

Red-eared Slider
p. 118

Painted Turtle
p. 118

Wood Turtle
p. 119

Common Snapping Turtle
p. 119

Leatherback
p. 119

Eastern Garter Snake
p. 120

Red-bellied Snake
p. 120

Smooth Green Snake
p. 121

Northern Ring-necked
Snake p. 121

FISH

Rainbow Trout
p. 124

Atlantic Salmon
p. 124

Brook Trout
p. 124

Lake Whitefish
p. 125

Rainbow Smelt
p. 125

Chain Pickerel
p. 125

FISH

White Sucker
p. 126

Emerald Shiner
p. 126

Banded Killifish
p. 126

Mottled Sculpin
p. 127

American Eel
p. 127

Sea Lamprey
p. 127

Atlantic Sturgeon
p. 128

Burbot
p. 128

Atlantic Cod
p. 128

Winter Flounder
p. 129

Atlantic Mackerel
p. 129

Atlantic Bluefin Tuna
p. 129

Alewife
p. 130

Thorny Skate
p. 130

Porbeagle Shark
p. 130

INVERTEBRATES

Acorn Barnacle
p. 133

Softshell Clam
p. 133

Blue Mussel
p. 133

American Oyster
p. 134

Atlantic Deep-sea Scallop
p. 134

Northern Moon Snail
p. 134

Common Periwinkle
p. 135

Green Sea Urchin
p. 135

INVERTEBRATES

Daisy Brittle Star
p. 135

Frilled Anemone
p. 136

Green Crab
p. 136

Northern Shrimp
p. 136

North American Lobster
p. 137

Moon Jellyfish
p. 137

Short-finned Squid
p. 137

Cabbage White
p. 138

Eastern Black Swallowtail
p. 138

Mourning Cloak
p. 138

American Copper
p. 139

Spring Azure
p. 139

Luna Moth
p. 139

Green Darner
p. 140

Common Whitetail
p. 140

Firefly
p. 140

Multicoloured Asian Ladybug
p. 141

Migratory Grasshopper
p. 141

Eastern Yellow Jacket
p. 141

Bumble Bee
p. 142

Horsefly
p. 142

Eastern Daddy Longlegs
p. 142

Balsam Fir
p. 146

White Spruce
p. 146

Black Spruce
p. 147

Red Spruce
p. 147

Eastern Hemlock
p. 148

Tamarack
p. 148

Eastern White Pine
p. 149

Jack Pine
p. 149

Red Pine
p. 150

Northern White-cedar
p. 150

American Elm
p. 151

American Beech
p. 151

Red Oak
p. 152

Yellow Birch
p. 152

White Birch
p. 153

Eastern Hop-hornbeam
p. 153

American Basswood
p. 154

Balsam Poplar
p. 154

Quaking Aspen
p. 155

Sugar Maple
p. 155

Red Maple
p. 156

White Ash
p. 156

SHRUBS & VINES

Pussy Willow
p. 159

Speckled Alder
p. 159

Witch-hazel
p. 159

Labrador-tea
p. 160

Bog Laurel
p. 160

Leatherleaf
p. 160

Highbush Blueberry
p. 161

Mayflower
p. 161

Wild Black Currant
p. 161

Pin Cherry
p. 162

American Mountain-ash
p. 162

Downy Serviceberry
p. 162

Steeplebush
p. 163

Rhodora
p. 163

Shrubby Cinquefoil
p. 163

Black Raspberry
p. 164

Black Chokeberry
p. 164

Swamp Rose
p. 164

Bunchberry
p. 165

Red-osier Dogwood
p. 165

Winterberry
p. 165

Staghorn Sumac
p. 166

Poison Ivy
p. 166

Buttonbush
p. 166

SHRUBS & VINES

Nannyberry
p. 167

Common Elderberry
p. 167

Bayberry
p. 167

Wild Cucumber
p. 168

Bittersweet
p. 168

Riverbank Grape
p. 168

HERBS, GRASSES, FERNS & SEAWEEDS

Canada Lily
p. 172

Painted Trillium
p. 172

Yellow Trout-lily
p. 172

False Solomon's-seal
p. 173

Canada Mayflower
p. 173

Bluebead Lily
p. 173

Indian Cucumber Root
p. 174

Sessile Bellwort
p. 174

Northern Blue Flag
p. 174

Arethusa
p. 175

Grass-pink
p. 175

Calypso
p. 175

Stemless Lady's-slipper
p. 176

Large Purple Fringed Orchid
p. 176

Early Coral Root
p. 176

HERBS, GRASSES, FERNS & SEAWEEDS

Wild Ginger
p. 177

Marsh-marigold
p. 177

Hepatica
p. 177

Bristly Buttercup
p. 178

Goldthread
p. 178

Red Baneberry
p. 178

Bloodroot
p. 179

Dutchman's Breeches
p. 179

Pokeweed
p. 179

Water Smartweed
p. 180

Pitcher-plant
p. 180

Shinleaf
p. 180

Common Wood Sorrel
p. 181

Narrow-leaved Sundew
p. 181

Blue Violet
p. 181

Sea Rocket
p. 182

Indian-pipe
p. 182

Fringed Loosestrife
p. 182

Starflower
p. 183

Sea-milkwort
p. 183

Purple Saxifrage
p. 183

Common Strawberry
p. 184

Silverweed
p. 184

Twinflower
p. 184

Lupine
p. 185

Beach Pea
p. 185

Eurasian Water-milfoil
p. 185

Purple Loosestrife
p. 186

Fireweed
p. 186

Common Evening-primrose
p. 186

Enchanter's-nightshade
p. 187

Beach Heath
p. 187

Spotted Touch-me-not
p. 187

Dwarf Ginseng
p. 188

Wild Sarsaparilla
p. 188

Snakeroot, Black Snake
p. 188

Smooth Sweet-cicely
p. 189

Water Parsnip
p. 189

Common Water-hemlock
p. 189

HERBS, GRASSES, FERNS & SEAWEEDS

Scotch Lovage
p. 190

Cow Parsnip
p. 190

Marsh-pennywort
p. 190

Angelica
p. 191

Indian-hemp
p. 191

Common Milkweed
p. 191

Bittersweet Nightshade
p. 192

Phlox
p. 192

Oysterleaf
p. 192

Blue Vervain
p. 193

Heal-all
p. 193

Wood Sage
p. 193

Common Monkeyflower
p. 194

Wood-betony
p. 194

Common Bladderwort
p. 194

Harebell
p. 195

Cardinal-flower
p. 195

Bluets
p. 195

Partridgeberry
p. 196

Northern Bedstraw
p. 196

Devil's Beggarticks
p. 196

Giant Ragweed
p. 197

Common Yarrow
p. 197

Oxeye Sunflower
p. 197

Common Tansy
p. 198

Golden Ragwort
p. 198

Canada Goldenrod
p. 198

New England Aster
p. 199

Common Boneset
p. 199

Canada Thistle
p. 199

Common Dandelion
p. 200

Broad-leaved Arrowhead
p. 200

Skunk Cabbage
p. 200

Jack-in-the-pulpit
p. 201

Lesser Duckweed
p. 201

Pickerelweed
p. 201

HERBS, GRASSES, FERNS & SEAWEEDS

Broad-leaved Cattail
p. 202

Beach Grass
p. 202

Common Reed Grass
p. 202

Rush
p. 203

Cord Grass
p. 203

Spike Grass
p. 203

Beach Sedge
p. 204

Cottongrass Bulrush
p. 204

Ostrich Fern
p. 204

Bracken Fern
p. 205

Sea Lavender
p. 205

Eel Grass
p. 205

Sugar Kelp
p. 206

Rockweed
p. 206

Encrusting Corralline Algae
p. 206

Dulse
p. 207

Irish Moss
p. 207

Sea Lettuce
p. 207

INTRODUCTION

Atlantic Canada has a personality unique from the rest of the country, with a dynamic culture, history and geography, and exceptional biodiversity. Despite being a small area of Canada with a relatively sparse population, an incredible cultural diversity exists, influenced by First Nations, Inuit, Innu and Métis peoples, along with the European colonial cultures of the English, Scottish, Irish and French. Europeans who visited but did not settle in the region include Basque whalers in the 1500s and the Vikings, who walked these lands over 1000 years ago. No less impressive is the incredible biodiversity of flora and fauna that flourishes here, from plant species such as the intricately designed pitcher-plant and the deliciously syrupy sugar maple to unique, iconic animals such as the Atlantic puffin and the lobster.

pitcher-plant

Often defined with different borders politically and biogeographically, in this guide, Atlantic Canada is treated as the large area that includes the three Maritime provinces of Nova Scotia, New Brunswick and Prince Edward Island, plus the island of Newfoundland, mainland Labrador, the mainland Atlantic coast of Québec, the St. Lawrence River and the Gaspé Peninsula, as well as numerous islands such as Québec's Magdalen Islands and Anticosti Island. We will discover nature all along the shores of the Gulf of St. Lawrence and the North Atlantic coast of Newfoundland and Labrador.

At the edge of the continent, worlds collide—terrestrial and marine, mainland and island, north and south, subarctic and temperate, sea and sky. Both the forests of the Maritimes and the waters of the Grand Banks are transitional ecosystems between two larger ecozones, and where two great

North Atlantic right whale

systems merge, biodiversity abounds. The **Atlantic Maritime Forest** is a unique transitional forest that includes elements of both the boreal forest and temperate broadleaf and mixed woodlands. The shallow waters of the Grand Banks mark the convergence of the cold, Arctic waters of the Labrador Current and the warm waters of the Gulf Stream.

Vertical transitions from great peaks high above sea level down to low valleys and estuaries also add to the diversity of ecosystems. In Canada, the northern Appalachian Mountains reach elevations of nearly 900 m in the higher peaks of the Long Range in Newfoundland, and the coastlines feature myriad fjords with sea cliffs rising to 300 m above sea level from which dramatic waterfalls plummet. These highlands contrast with the warm lowland forests and sheltered bays, which have the warmest ocean waters in Canada.

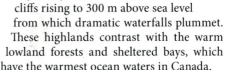

sharp-shinned hawk

horsefly

Parts of Atlantic Canada are busy with human endeavours—industrial, agricultural and urban—yet are still wild enough for foxes, bears and hawks; other areas are remote enough for rare and elusive species such as bats, polar bears and wolverines. Our own backyards host visits from bold and opportunistic creatures such as coyotes, deer and many species of birds, insects and rodents. And the region features world strongholds for populations of woodland caribou and beluga whales. We get to enjoy the great spectacle of migratory birds travelling along the Atlantic Flyway in spring and autumn, and see coastal cliffs completely covered in millions of flapping, squawking, nesting seabirds. We can wrap our arms around ancient trees in our great forests and listen to choruses of frogs singing from the wetlands, lakes and rivers. There is much to appreciate and admire in Atlantic Canada.

northern spring peeper

woodland caribou

NATURAL REGIONS

Boreal-Taiga Forests

The boreal forest, which includes the taiga (boreal and taiga are synonyms outside Canada), is the world's largest biome, a circumpolar forest of which one-third lies in Canada. The **Eastern Canadian Shield Taiga** is an ecoregion of boreal forest that is found in northern Québec and most of Labrador. The domi-

polar bear

nant vegetation here is coniferous trees, namely spruce, fir and pine. Wolves, martens and wolverines are the elusive inhabitants of

black spruce

the northern boreal forest. At low elevations, deciduous woodlands form along the many rivers, streams and wetlands that are habitat for many species of fish and amphibians, as well as mammals such as moose and beaver. Above this heavily forested region are alpine tundra and high mountain ranges and plateaus with lakes and string bogs. The trees peter out at what is called the timberline or tree line, where no forests grow. Here, the boreal forest becomes taiga dwarf forest—a zone in which normally straight and tall trees grow gnarled, twisted and stunted from the forces of the harsh climate. This type of forest is also known as "krummholz."

In the far northern tundra, wolves tirelessly follow the caribou herds, which in turn feed almost exclusively on tiny lichens. Precipitation is low and occurs mainly in the form of snow—winters are long and summers are but a brief blip. The ground stays permanently frozen to depths of one metre and is called permafrost. Only the most tenacious plants such as purple saxifrage grow here.

The north coast of Labrador is the gateway to the Arctic. These shores host charismatic Arctic mammals—polar bears prowl the shoreline and harp and hooded seals give birth to their young on the ice floes in spring. Arctic foxes are often seen following polar bears in the hope of scavenging a meal. The rocky coastline features barnacles, limpets, mussels, crabs, sea anemones and other marine invertebrates, and the cold waters are preferred by the large brown seaweeds, such as wrack and kelp, but a few of the red and green seaweeds can also be seen.

frilled anemone

The Torngat Mountains, found at the extreme northern tip of Labrador and eastern Québec, are the southernmost point of the **Arctic Cordillera**. These barren, glaciated mountains are the highest peaks in mainland Canada east of the Rockies, and the tallest is Mount Caubvick (Mont D'Iberville in Québec) at 1652 m elevation. The ranges are separated by deep fjords and finger lakes surrounded by sheer rock walls.

The **Eastern Canadian Boreal Forest** is influenced by the warming maritime climate. The forest is still dominated by conifers, mainly black and white spruces, but balsam fir is the climax species; white birch and aspen are typical of disturbed sites. The understorey is made up of mosses and bog laurel, forming an almost continuous carpet except on the higher, more exposed peaks, where krummholz of twisted spruce and fir occur and barrens appear. Barrens are highlands and northern exposed zones, a hostile environment where only the hardiest of plants can grow. The South Avalon–Burin Oceanic Barrens on the Avalon Peninsula of Newfoundland is an example. Mosses, heaths, lichens, shrubs, bogs and dwarf trees are characteristic. Willow ptarmigan and endangered woodland

willow ptarmigan

caribou occur inland. Granite outcroppings and exposed bedrock where glaciers scraped over the land provide stark contrast to the dark, dense forests. This eco-type is found in eastern Québec, including higher elevations on the Gaspé Peninsula, as well as the highlands and eastern exposed coasts of Newfoundland, New Brunswick and the Cape Breton.

Temperate Broadleaf and Mixed Forest

The **Maritime Transitional Forest** is unique and dynamic—the forest has no rigid composition and is almost chameleon-like in nature. In some areas, it is what you expect of the classic **Acadian–New England Forest** of red oak, sugar maple, yellow birch and beech; yet, in other parts, it appears boreal, with bogs and stands of black spruce or balsam fir. Indeed, it is a little bit of both of these forest types blended together, overlapping, appearing and disappearing throughout the region. The transition zone occurs between the boreal spruce-fir forest to the north and the deciduous forest to the south, with the Atlantic Ocean strongly influencing vegetation dynamics in coastal areas.

Little of the original mixed forest remains, and much of the old-growth sugar maple, American beech, American elm, black ash, yellow birch, white pine and eastern hemlock were cleared for farms or by logging, fires, insect infestations and hurricanes. White spruce, black spruce, balsam fir and tamarack predominate today. Old-growth stands characterized by red oak, sugar maple, yellow birch and beech still exist at higher elevations. Bogs and fens are common. Moisture permits rich beds of moss and lichen to grow on tree trunks and the gnarled roots of beech trees. American elm, black ash and red maple are typical of wet sites.

sugar maple

In the **Gulf of St. Lawrence Lowlands**, the scent of balsam fir and balsam poplar fills the air. Also within these lowland forests is eastern hemlock, an evergreen found in deciduous forests. Its fallen needles on the forest floor create acidic soils that encourage heaths to grow, especially blueberries. American elm, black ash, white pine, red maple, red oak and black, red and white spruce can also be found. The lowland forest covers all of PEI, the Magdalen Islands, most of east-central New Brunswick, the Annapolis Valley, Minas Basin and the

red fox

Northumberland Strait coast of Nova Scotia. Black bears, red foxes, fishers, beavers, raccoons, white-tailed beer, snowshoe hares, muskrats and porcupines inhabit the temperate broadleaf and mixed forest region. Wolves were eradicated over a century ago, and coyotes have taken their place.

The Rocky Shore

The rocky shore is divided into the **spray zone** and the **intertidal zone**. The spray zone lies just beyond the reach of the highest high tides, leaving the vegetation to receive moisture in the form of sea spray or from high, stormy waves or rainfall. Eastern Newfoundland and the foggy, cool Bay of Fundy has classic rugged, rocky shores, exposed to high winds and continuous salt spray. White spruce is a dominant tree species in coastal areas owing to its tolerance of sea salt spray. Barnacles and periwinkles cover the rocks.

*common
periwinkle*

blue mussel

The **intertidal zone** lies between the high and low tide lines and has several subzones. The **upper intertidal zone**, also known as the periwinkle zone, is near the high tide line, and animals and algae here must survive almost half the day without water. Shellfish, such as blue mussels, must trap sea water or close off fresh water as needed. Periwinkles and dog whelks (*Nucella lapillus*) graze on the thin film of green algae on the rocks and in tidal pools. The **middle intertidal zone**, also called the barnacle-rockweed zone, forms the largest part of the inter-

tidal zone. Characteristic flora and fauna are brown algae, predominantly rockweed (*Fucus* spp.) and knotted wrack (*Ascophyllum nodosum*), periwinkles, barnacles, blue mussels, limpets (*Acmaea* spp.) and sea stars. The **lower intertidal zone**, or Irish moss zone, lies just above the low tide mark. Irish moss, a type of red algae, is only exposed during the very lowest tides. There is higher animal diversity here

Irish moss

owing to the stability of the habitat—extreme tidal variations are infrequent, making life in this zone less stressful to organisms. Blue mussels are quite abundant here. The **subtidal zone**, or kelp zone, is always underwater except during the extreme spring tides of winter. This zone extends only as far as sunlight can penetrate, and it is inhabited by leathery kelp and red algae, particularly encrusting coralline algae, as well as sea stars and sea urchins. Flounders and eels enter the intertidal zone at high tide.

Sandy Beaches and Wrack Lines

The **sandy beach** is a zone of constantly shifting unconsolidated material—sand. Created by riparian deposits, coastal erosion or glacial till, the grains of sand are diverse in colour. If you look closely, you will notice black, green, red and brown grains from various origins, as well as white or clear grains, which are quartz. The mineral grains dissolve over time,

raccoon

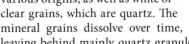

leaving behind mainly quartz granules; thus, older sand is less colourful, and white sandy beaches are old beaches. Also, the coarser the sand, the steeper the face of the beach.

beach grass

Dead seaweed and other flotsam collect at the **wrack lines**, attracting little beach fleas or beach hoppers, as well as many other invertebrates, providing food for gulls, plovers, sanderlings, terns, sandpipers, raccoons, skunks and various rodents. Pioneer plants on the dunes, a few metres landward and above the reach of the highest storm waves, include beach grass, dusty miller (*Artemisia stelleriana*) and beach pea. The next succession of vegetation grows on the soils stabilized by the pioneer plants and includes coastal species of wild roses (*Rosa* spp.) and goldenrod (*Solidago* spp.), bayberry, poison ivy and coastal grasses.

dunlin

Salt Marshes

Salt marshes occur in protected bays and estuaries where fresh water flows to the coast and into the ocean—the mixed fresh and salt water is known as brackish water. Tidal flooding and the stream currents deposit significant amounts of minerals and sediment in the shallow waters, forming the base of the marsh. Plants begin to take hold and stabilize the sedimentary floor, then decomposing plant materials start building layers of peat. Salt marshes are one of most highly productive ecosystems

staghorn sumac

sea lavender

owing to the nutrient-rich detritus they produce, and they are often referred to as the "nurseries of the sea" because many young fish and crustaceans can be found in these sheltered, food-rich waters. Many species of shellfish, fish and shorebirds rely on saltmarsh habitat. The highest biodiversity occurs at the marsh edge farthest from the ocean tides, starting with plants such as large reeds, staghorn sumac and bayberry.

Salt marshes can typically be divided into three zones. The **upper marsh** is a freshwater zone with freshwater plants, but they are salt tolerant and are submerged during spring and storm tides. These plant species include spike grass, rushes, sea lavender, goldenrod and asters. The **middle marsh**, or salt meadow, is almost dominated by cord grass (*Spartina* spp.). In the **lower marsh**, closest to the ocean, cord grass is often submerged. Green algae and small snails are abundant, as is the saltmarsh mosquito.

sea lettuce

Tidal Flats

softshell clam

When the lowest tides are out, we can see—and smell—the mud and sand of the low-wave situations in estuaries. These exposed flat areas are called **tidal flats**. The sediment constantly shifts, so most plants cannot take hold, but free-floating algae—diatoms and dinoflagellates—cover the sand in a greenish yellow sheen, and eel grass is prevalent. Little clams and sea worms dig into the sand—you can see the little breathing holes they create, and crabs, small crustaceans, snails and other molluscs are found throughout the flats.

The St. Lawrence River and the Gulf of St. Lawrence

The Gulf of St. Lawrence is one of the world's largest estuaries and is fed by the St. Lawrence River—the second largest river in Canada—and its tributaries. An amazing diversity of life that moves between saline and fresh water occurs at the confluence of the St. Lawrence River and the gulf. Atlantic cod are found in the deeper parts of the river, whereas smelt inhabit the entire estuary. Beluga whales, harbour porpoises and harbour seals move between the river and the estuary. Atlantic salmon and brook trout are migratory fish. Clams, mussels, sponges and sea stars inhabit the benthos of the river bottom.

daisy brittle star

31

The Atlantic Ocean

Farther from the coast, the warmer marine waters from the Gulf of Maine to the Davis Strait meet the Arctic-chilled waters of the deep Northwest Atlantic. Cold, saline waters from the Labrador Sea flow along the western edge of the North Atlantic, creating the Labrador Current. The warm Gulf Stream

Atlantic cod

current moves northward along this western edge. The two currents collide and are redirected east over the Grand Banks, a group of shallow (24 to 100 m deep) underwater plateaus. The mixing of the warm and cold currents over these shallow banks lifts nutrients off the ocean floor and creates one of the richest fishing grounds in the world. The Atlantic cod, an iconic Grand Banks species, is now recuperating from overharvesting, its near disappearance an inconceivable occurrence for anyone who fished for cod

herring gull

when it existed in epic numbers. Many other fish species swim these waters, and crustaceans—lobster, crabs and shrimp—are abundant. There are many species of seabirds and marine mammals, and birding and whale watching are hugely popular and rewarding along the Atlantic coast.

North American lobster

The meeting of cold and warm water currents makes for extremely foggy conditions. Icebergs, some larger than cruise ships, are abundant between Newfoundland and Greenland, the infamous Iceberg Alley where the *Titanic* and numerous other ships have been lost in the ice and fog. Thousands of icebergs break off the ancient glaciers in Greenland and are carried to Canada on the Labrador Current during the spring iceberg season, sometimes floating as far south as Nova Scotia.

Atlantic deep-sea scallop

blue whale

HUMAN-ALTERED LANDSCAPES & URBAN ENVIRONMENTS

The impact of human activity on natural environments is something we must become increasingly aware of and sensitive to as our populations continue to encroach on wildlife habitat. No description of important habitats would be complete without a mention of towns and cities. Roads, urban and agricultural areas and forestry and mining sites are just a few examples of the impact that humans have on the landscape. The Maritime and Gaspé regions have lost much of their original forest to forestry and agriculture; the large, ancient trees were valuable timber and textile for early homesteaders, and the rich soil produced bountiful fruit orchards and hay and vegetable crops. The St. Lawrence Lowlands region supports ranching and dairy farms, which supply butter, cheese and meat for the region. But these activities have had a significant impact on the wildlife of the area. Black bears, fishers, bobcats and martens have been extirpated from PEI, and wolves, never present on PEI, have been exterminated from the island of Newfoundland, as well as New Brunswick and Nova Scotia.

eastern white pine

Biodiversity is at its highest along the suburban fringe, where a botanical anarchy of remnant native plants, exotic introduced plants and hybrids exists. Strategic species, whether native or introduced, take advantage of evolving opportunities for food, shelter and breeding territory. We have established human-made lakes, urban parks, bird feeders, birdhouses and bat houses to deliberately accommodate the species we appreciate, whereas wharves and ports, garbage dumps and even our own homes seem to attract unwanted species we consider to be pests.

eastern red bat

house mouse

Many of the most common plants and animals in these altered landscapes were not present before the arrival of settlers and modern transportation. The most established of the introduced species exemplify how co-habitation with humans offers a distinct set of living situations that benefit many plants and animals. The house mouse, Norway rat and house sparrow are some of the highly successful exotic animals that have been introduced to North America from Europe and Asia.

house sparrow

33

THE SEASONS

The seasons of Atlantic Canada greatly influence the lives of plants and animals. Although some birds, insects and marine mammals are migratory, most animals are terrestrial and have varying ranges—the range of a wolf can cover several hundreds of hectares, and a whale can travel the length of the North American continent, but a vole's

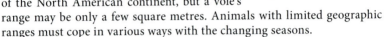

leatherback

range may be only a few square metres. Animals with limited geographic ranges must cope in various ways with the changing seasons.

meadow vole

With rising temperatures, reduced snow or rain and the greening of the landscape, spring brings renewal. Many animals bear their young at this time of year. An abundance of food travels through the food chain—lush, new plant growth provides ample food for herbivores, and the numerous herbivore young become easy prey for carnivores. Whereas some small mammals, particularly rodents, mature within weeks, the offspring of large mammals depend on their parents for much longer periods.

During summer, animals have recovered from the strain of the previous winter's food scarcity and spring's reproductive efforts, but it is not a time of relaxation. To prepare yet again for the upcoming autumn and winter, some animals must eat vast quantities of food to build up fat reserves, whereas others work furiously to stockpile food caches in safe places. Some of the more charismatic species such as white-tailed deer and moose mate in the autumn with dramatic ruts, and some small mammals such as voles and mice mate every few months or even year-round.

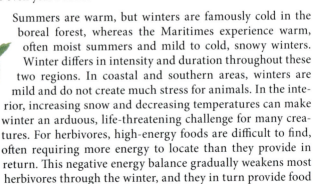

moose

Summers are warm, but winters are famously cold in the boreal forest, whereas the Maritimes experience warm, often moist summers and mild to cold, snowy winters. Winter differs in intensity and duration throughout these two regions. In coastal and southern areas, winters are mild and do not create much stress for animals. In the interior, increasing snow and decreasing temperatures can make winter an arduous, life-threatening challenge for many creatures. For herbivores, high-energy foods are difficult to find, often requiring more energy to locate than they provide in return. This negative energy balance gradually weakens most herbivores through the winter, and they in turn provide food

black raspberry

American redstart

for the equally needy carnivores. Voles and mice find advantages in the season—an insulating layer of snow protects their elaborate trails from the worst of winter's cold. Food, shelter and warmth are all found in the thin layer between the snow and the ground surface, and the months devoted to food storage now pay off.

The seasons also affect the array of species found in the region. When you visit natural areas in winter, for example, you will see a different group of species than in summer; many plants die back, migrating animals head south, and other animals become dormant in winter. Conversely, many birds arrive at bird feeders in winter, and certain mammals, such as deer, enter lowland meadows to find edible vegetation, making these species more visible during cold weather.

northern moon snail

Spring tides are the highest and the lowest when the moon and the sun (both on the same side of the planet, the moon in front of sun) are aligned with the earth, generating the maximum gravitational pull; neap tides are the most minimal tidal changes, when the sun and the moon are at right angles to each other, cancelling out one another's gravitational pull on the oceans. The moon orbits the earth every 27.5 days, changing position daily and rising about 50 minutes later each day. So, at the same point on the earth, tidal changes occur 50 minutes later each day. This explains one tide change every day, but most places have two high and two low tides daily. The second tide is a result of centrifugal force, a whipping effect that occurs as the earth turns.

thorny skate

No discussion about the weather and the seasons would be complete without mentioning hurricanes. Atlantic Canada is not immune to the impacts of tropical cyclones and has been hit by several severe hurricanes throughout recorded history. Hurricane Igor in the autumn of 2010 is the most recent example, and one of the earliest on record occurred in 1775, when the Newfoundland Hurricane took the lives of some 4000 mariners, mostly British; it is considered by some to be the critical factor in the defeat of the British in the American Revolution. Between then and now, hurricanes have hit the coast of Atlantic Canada with random frequency.

white-tailed deer

NATIONAL PARKS, PROTECTED AREAS & OTHER WILDLIFE-WATCHING AREAS

Whether with great foresight or regretful hind-sight, many protected areas and provincial and national parks have been established over the recent few decades to conserve areas of wilderness that were at risk of being lost. Remote location and sparse population has saved other areas, leaving great tracts of wilderness in the north virtually untouched. During the early decades of European settlement, species disappeared because of either habitat loss or deliberate extirpation—the great auk, the Labrador duck and several races of Maritime

bunchberry

wolves, such as the Labrador and Beothuk wolves. Ancient old-growth forests were wiped out because wooded areas were cleared for pulp and timber or for agriculture, and predator eradication was implemented to prevent livestock losses. Today, an ever-increasing acreage of Atlantic Canada is being set aside for wilderness conservation, for its intrinsic value as well as for us to appreci-ate and experience today and into the future. Below is just a sampling of the parks and natural areas you may wish to explore.

Fundy National Park, New Brunswick

This park protects some of the last wilderness in southern New Brunswick and two important environmental sys-tems: the lush, conifer-dominated Caledonia Highlands, which look like mystical lands in the clouds, and the foggy Bay of Fundy below. The tidal range in the bay is about 5.8 to 6.7 m (depending on the phase of the moon) and is created by ocean water being funnelled into a narrow, contained bay. The tides become progressively higher moving northeast along the Bay of Fundy toward St. John. Minas Basin in Nova Scotia has had world-record high tides. The coastal system comprises rocky shores, tidal flats, salt marshes and seaside cliffs. The Caledonia Highlands, 300 m above the coast, is Acadian forest, with red spruce,

pileated woodpecker

balsam fir, birches, maples and bunchberry, as well as plenty of bogs that host pitcher-plants. Beavers, moose, flying squirrels, pile-ated woodpeckers, warblers and green frogs are just a few of the animals to be seen.

green frog

Top Wildlife-watching Sites in Atlantic Canada

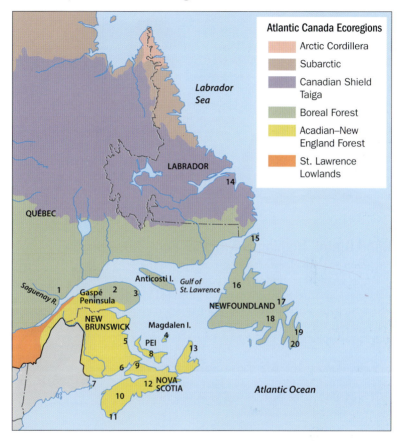

Atlantic Canada Ecoregions
- Arctic Cordillera
- Subarctic
- Canadian Shield Taiga
- Boreal Forest
- Acadian–New England Forest
- St. Lawrence Lowlands

Maritime Québec (QC)
1. Saguenay–St. Lawrence Marine Park
2. Gaspésie National Park
   (*Parc national de la Gaspésie*)
3. Forillon National Park
4. Magdalen Islands
   (*Îles de la Madeleine*)

New Brunswick (NB)
5. Kouchibouguac National Park
6. Fundy National Park
7. Grand Manan Island

Prince Edward Island (PEI)
8. Prince Edward Island National Park

Nova Scotia (NS)
9. Amherst Point Bird Sanctuary
10. Kejimkujik National Park
11. Cape Sable Island
12. Shubenecadie Provincial Wildlife Park
13. Cape Breton Highlands National Park

Newfoundland and Labrador (NL)
14. Mealy Mountains National Park Reserve (proposed)
15. Burnt Cape Ecological Reserve
16. Gros Morne National Park
17. Terra Nova National Park
18. Bay du Nord Wilderness Reserve
19. Witless Bay
20. Avalon Wilderness Reserve

Kouchibouguac National Park, New Brunswick

In 1968, 238 km² of Maritime plains were set aside to establish this national park, and since that time, it has protected the second largest tern colony in North America, as well as rare sand dune ecosystems. Colonies of harbour and grey seals can be found on

grey seal

red-backed salamander

some of these coastal dune shores. Bogs, salt marshes, freshwater streams and lagoons provide habitat for almost all the amphibian species that inhabit eastern Canada. Kouchibouguac is Mi'kmaq for "river of the long tides" and is named after the tidal rivers that flow through this park. Indeed, it is a wonderful place to canoe or kayak.

Cape Breton Highlands National Park, Nova Scotia

Highland plateaus elevated on steep cliffs above the Atlantic Ocean are the most striking feature of northern Cape Breton and have been protected in this 950 km² park since 1936. One-third of the famous Cabot Trail runs through the park. Acadian, boreal and taiga habitats can all be found here, creating an assemblage of flora and fauna not seen elsewhere in Canada. Among the 631 native plants, both sugar maple and balsam fir occur, and lynx and snowshoe hare are found in the park, as well as black guillemots, cormorants and other seabirds.

black guillemot

Kejimkujik National Park, Nova Scotia

In a region characterized by its relationship to the ocean, sometimes the inland wilderness areas are overlooked. Not so in Kejimkujik. Here, not only are the coastal estuaries and salt-marsh habitats protected, but also the uplands. You can hike many of the beautiful, rolling woodland trails through several forest types, including old-growth hemlock stands, or you can explore historic canoe routes and portages. No river trip is complete without seeing a beaver dam, or perhaps even a beaver, and while paddling on the lake, you might spot another wilderness icon, the common loon. Deer and porcupines are also possible to see, as are any of the 178 bird species that occur here.

eastern hemlock

porcupine

Amherst Point Bird Sanctuary, Nova Scotia

This nationally protected 1000 ha habitat is a haven for approximately 228 bird species. Not only a paradise for birders, the sanctuary is also a beautiful area in which to hike. Visitors can walk the 8 km of trails, 2.5 km of which is an interpretive trail around Layton's Lake, and take in the scenic old-growth forests of fir, aspen, birch, spruce and maple, which are stunning in the fall. Resident mammal species include red squirrels, snowshoe hares and muskrats.

red squirrel

Prince Edward Island National Park, PEI

The Maritime plain that is protected within this park is characterized by sand dunes, barrier islands, sandspits and wetlands. Beach grass is one of the most important plants on the beaches, stabilizing dunes and allowing other plants to take hold. In 1998, in order to protect the rare dune ecology that lay beyond the park boundary, 6 km of the Greenwich Peninsula were added to ensure their conservation, as well as the future of 10,000-year-old archaeological sites. The park also cherishes Green Gables, part of L.M. Montgomery's Cavendish National Historic Site, and the Dalvay-by-the-Sea National Historic Site. The park has been designated an Important Bird Area, and the protected beaches provide nesting for the endangered piping plover (*Charadrius melodus*).

beach heath

marten

Terra Nova National Park, Newfoundland

The brunt of the North Atlantic's forceful waves pounds the rocky headlands of Newfoundland. The rugged cliffs contrast with the sheltered inlets and the inland forests and bogs. The wildlife in this park is exciting to discover and includes salmon, black bears and the endangered Newfoundland marten. The natural and cultural significance of the area was obvious back in 1957, when this park was established as Newfoundland's first national park.

Gros Morne National Park, Newfoundland

Gros Morne National Park was designated a UNESCO World Heritage Site in 1987 owing to its rich cultural fishing history set in a unique and beautiful ecological area. Here, you can explore the uninhabited Long Range Mountains or some the many fishing villages. Camp on a sandy beach full of shellfish or take a boat tour to view the towering cliffs in the freshwater fjords carved by glaciers. Along the coast you can see forests of stunted, gnarled balsam fir and spruce trees, a feature known locally as "tuckamore." The park's wildlife includes caribou, lynx, black bears, martens and introduced moose, and the surrounding seas support populations of marine mammals and sea ducks, including the harlequin duck (*Histrionicus histrionicus*).

lynx

Mealy Mountain National Park Reserve, Labrador

Although not yet officially a national park, the proposed Mealy Mountain National Park Reserve would encompass 10,700 km², making it the largest national park in Atlantic Canada. The area has remained wild, natural and undeveloped without official protection because of

brook trout

its remote location. The Mealy Mountains rise to heights of over 1000 m from the shores of Lake Melville in south-eastern Labrador, creating islands of Arctic tundra within the boreal forests. Boreal species such as moose, woodland caribou, black bears and ospreys are at home here. Glacier lakes and wetlands are important habitat for the migratory birds that breed or rest along the Atlantic Flyway, as well as for salmon and brook trout.

osprey

Gaspésie National Park, Québec

This 800 km² inland park at the end of the Gaspé Peninsula is home to the only caribou herd south of the St. Lawrence River. High Appalachian peaks offer alpine tundra in contrast to the St. Lawrence estuary, where beluga whales frolic.

beluga whale

Saguenay–St. Lawrence Marine Park, Québec

The federal and provincial governments created this marine park under joint management in 1998 to protect the marine environment of the 1245 km² area at the confluence of the St. Lawrence Estuary and the Saguenay Fjord. Beluga whales, harbour porpoises and three other cetacean species now have their habitat protected to the best degree currently possible while the St. Lawrence Seaway remains an important commercial shipping lane. The adjacent provincial parks are Saguenay Fjord National Park (*Parc national du Fjord-du-Saguenay*) and Saguenay National Park (*Parc national du Saguenay*). Saguenay Fjord National Park protects 100 km of the bays, coves and sea cliffs that comprise the mouth of the fjord, whereas Saguenay National Park comprises the north shore of the St. Lawrence River for 90 km northeast and southwest of the Saguenay River. The riverbeds are at depths of 300 m, and tides bring cold, salty waters from the Gulf of St. Lawrence into the fresh water at the conflux of the two rivers. This mixing and layering of fresh and saline, warm and cold, makes for enriched plant and animal diversity in the basins of the fjord.

sugar kelp

Forillon National Park, Québec

Created in 1970, Forillon National Park comprises 244 km² of coastal mountains at the far end of the Gaspé Peninsula and is representative of the Notre-Dame and Mégantic mountain regions. Rock formations and dramatic sea cliffs have been carved by the relentless crashing of waves. At the tip of the Gaspé Peninsula, the mountains meet the sea in a series of spectacular, multihued cliffs and plunging headlands. Seabird colonies nest on the cliffs and rare Arctic alpine plants can be found at high elevations. The Grande-Grave National Heritage Site within the park illustrates the way of life of fishing families.

double-crested cormorant

Wilderness Reserves

Wilderness reserves can be found throughout the region. In Newfoundland, the Bay du Nord Wilderness Reserve (2895 km²) is that province's largest reserve and one of North America's unspoiled barrens, hosting the winter calving grounds of some 15,000 caribou and habitat for a large Canada goose population. Brook trout and salmon are found in the Bay du Nord River, and the waterway provides excellent fishing opportunities for anglers, as well as river canoeing that is a true wilderness experience.

Canada goose

The Avalon Wilderness Reserve (1070 km²) of barrens and forests is home to the Avalon woodland caribou herd, the most southerly caribou herd in Canada. The glacier-carved landscape of barrens, ponds, rivers, bogs, small forests and thickets provide habitat for species such as willow ptarmigan, pitcher-plants and lichens, a staple of the caribou's diet. Six important salmon rivers run through the reserve.

Atlantic salmon

Ecological Reserves

Ecological reserves, which are typically less than 1000 km² in size, protect unique, rare or endangered flora and/or fauna (typically seabirds, such as the Atlantic puffin), as well as fossils. The areas are established for scientific research or educational purposes, and though some will permit a certain level of restricted, low-impact recreational activity, others are strictly off limits to all but scientists conducting research. One example is Funk Island, one of the historic nesting areas of the now-extinct great auk. It is now an important nesting area for many other seabirds, and access is only permitted for research purposes.

Atlantic puffin

Another reserve is Burnt Cape in Newfoundland, where rare plants are being studied in relation to the effects of climate change. Burnt Cape has also been known as "penguin island," in reference to the flightless auk's similarity to the penguin. Witless Bay Ecological Reserve, just off the east coast of the Avalon Peninsula, is a 31 km² reserve (29 km² is the marine component) that protects the largest Atlantic puffin colony in North America—more than 260,000 pairs of Newfoundland's official bird nest here.

American copper

Islands

Numerous islands throughout the Bay of Fundy are wild-life oases. One example is Grand Manan Island in New Brunswick. Although introductions of exotic animals and the logging of beech, oak and pine trees when settlers first arrived in the late 1700s has changed the original ecology, there are nevertheless still forests of spruce, fir, tamarack, birch and maple, and bogs full of native orchids and insectivorous plants.

yellow birch

Since 1801, Sable Island, Nova Scotia, has been a restricted island that people may visit only with permission. The restriction was originally implemented to stop the plundering and pilfering from shipwrecks, and today, this same restriction protects the fragile island dune ecology and the remaining 40 to 50 wild herds of approximately 200 Sable Island horses that roam the dunes. These feral horses are descended from livestock that Thomas Hancock of Boston sent to the island in 1760.

Shubenacadie Wildlife Park in Nova Scotia provides 0.4 km² of habitat for relocated Sable Island horses and other wildlife that is unable to be returned to its natural wild habitat.

tamarack

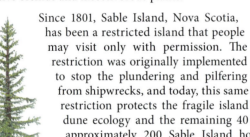

eastern garter snake

*stemless
lady's-slipper*

OBSERVING NATURE

Nature is always around us in some form, and not a day goes by when we do not see a plant or animal or notice the weather. In Atlantic Canada, the species and scenery are so abundantly gorgeous that we risk spoiling our senses. Wild beauty is ever present—even our neighbourhoods and backyards host numerous species of plants and animals. Consider that city parks offer ample opportunity to revel in nature, listen to songbirds or the buzzing of insects, smell the perfume of flowers or the decadence of autumn ripeness, see the colour of plants and birds, and watch the ongoing dramas and occupations of busy squirrels and courting nesting birds. On the coast, you can watch seabirds dive for fish or perhaps glimpse a humpback whale breech.

humpback whale

bumble bee

The Best Viewing Times

Many birds and mammals are most active at dawn and at dusk, so the best times for viewing them are during these "wildlife hours" when animals emerge from their daytime hideouts or roosting sites. During winter, hunger may force some mammals to be more active during midday. Conversely, in warm seasons, some animals may become less active and less visible in the heat of the day.

eastern pipistrelle

Birding Basics

The Atlantic Flyway is an important migratory route, bringing birds to our area that are travelling between the Arctic and their wintering grounds to the south. Birding is an increasingly popular activity for many people, and there are scores of excellent books, online resources, clubs and organizations to learn from. When practised with patience and reserve, birding is a low-impact pastime and a ready source of mental and physical exercise. One must be patient, though, because birds are among the most highly mobile animals; they may be seen one moment and then vanish the next!

veery

sora

Atlantic Canada hosts many breeding birds and year-round residents, as well as large numbers of spring and autumn migrants. Approximately 320 bird species can be seen in the region on a regular basis. Many pelagic seabirds, which only come to nest, do so on our coastlines in summer. These same coastlines are wintering grounds for waterfowl, geese, loons and grebes. Some of the largest common murre (*Uria aalge*) breeding colonies lie off our rocky coasts, and Atlantic puffins are found nowhere else in Canada. The Gannet Islands, off the northeast coast of Labrador, constitute the largest and most diverse seabird colony in Labrador.

northern gannet

Even in winter, the mild climate of the Maritimes permits many bird species to overwinter along the coast, and Christmas bird counts tally up scores of species. Some people set up bird feeders in their backyards in winter and stock them through to late spring to help birds before the plants bloom or insects hatch. Be sure to keep bird feeders clean, especially nectar feeders, to prevent the spread of disease or spoilage that can make birds ill.

pied-billed grebe

Naturescaping

Native plants in gardens and landscaping provide natural foods and shelter for birds, as well as for beneficial insects such as butterflies and bees, and even certain mammals. Flocks of waxwings have a keen eye for red mountain-ash berries, and hummingbirds enjoy columbine flowers. The cumulative effects of "naturescaping" in urban yards can be a significant step toward habitat conservation (especially when you consider that habitat is often lost in small amounts—a seismic line is cut in one area or a highway is built in another). Many good books and websites about attracting wildlife to your backyard are available.

cedar waxwing

cardinal-flower

In contrast, know how to keep your yard from attracting easily habituated wildlife such as coyotes and bears. Deal with garbage and compost responsibly, reconsider bird feeders that may lure bears, and do not leave pet food outside. When camping, be bear-wise with your food and toiletries, storing them in a manner that will not attract the keen noses and insatiable appetites of these animals.

eastern black swallowtail

Whale Watching

Whale watching can be an organized activity, with boats taking groups of tourists out to known areas of high whale and dolphin sightings, but because many aquatic species frequent inshore waters,

Atlantic white-sided dolphin

there can be random moments of fortune when you might see a whale or dolphin right from the shore. Although whale watching has strong merit for encouraging public awareness of and appreciation for marine mammals and the health of the oceans, it can disrupt cetacean behaviour, so tour groups must be considerate and passive in the presence of these sensitive species. Encourage companies and fellow whale watchers to not harass the animals when trying get as close as possible for a photo op, and understand the decisions of tour organizers to keep a respectful distance.

Humans and Wildlife

Although more people have become conscious of the need to protect wildlife, human pressures have nevertheless damaged critical habitats, and some species experience frequent harassment. Modern wildlife viewing demands courtesy and common sense. Honour both the encounter and the animal by demonstrating a respect appropriate to the occasion. Here are some points to remember for ethical wildlife watching in the field:

- Stress is harmful to wildlife, so never chase or flush animals from cover or try to catch or touch them. Use binoculars and keep a respectful distance, for the animal's sake and often for your own. Amphibians are especially sensitive to being touched or held—sunscreen or insect repellent on your skin can poison the animal.

blue-spotted salamander

- Leave the environment, including both flora and fauna, unchanged by your visit. Tread lightly and take home only pictures and memories. Do not pick wildflowers, and do not collect sea stars, sea urchins or seashells still occupied by the animal.

- Fishing is a great way to get in touch with nature, and many anglers appreciate the non-consumptive ethos of catch-and-release.

burbot

- Pets hinder wildlife viewing and may chase, injure or kill other animals, so control your pets or leave them at home.

- Take time to learn about wildlife and the behaviour and sensitivity of each species.

ANIMALS

Animals are mammals, birds, reptiles, amphibians, fish and invertebrates, all of which belong to the Kingdom Animalia. They obtain energy by ingesting food that they hunt or gather. Mammals and birds are endothermic, meaning that their body temperature is internally regulated and will stay nearly constant regardless of the surrounding environmental temperature,

red squirrel

unless the external temperature is extreme and persistent. Reptiles, amphibians, fish and invertebrates are ectothermic, meaning that they do not have the ability to regulate their own

razorbill

internal body temperature and tend to be the same temperature as their surroundings. Animals reproduce sexually, and they have a limited growth that is reached at sexual maturity. They also have

spotted salamander

diverse and complicated behaviours that are displayed in courtship, defence, parenting, playing, fighting, eating and hunting, as well as how they establish and recognize social hierarchies and how they deal with

Atlantic cod

environmental stresses such as weather, change of season or availability of food and water. This guide includes the region's most common, wideranging, charismatic and historically significant animals. Diverse families such as rodents are represented by a few selected species.

green darner

MAMMALS

Mammals are the group to which human beings belong. In general, mammals are endothermic, bear live young (with the exception of the platypus), nurse their young and have hair or fur on their bodies. Typically, all mammals larger than rodents are sexually dimorphic, meaning that the male and the female differ in appearance, either by size or by other diagnostics such as antlers. Males are usually larger than females. Different groups of mammals include herbivores, carnivores, omnivores and insectivores. People often associate large mammals with wilderness, making these animals prominent symbols in Native legends and stirring emotional connections with people in modern times.

Whales, Dolphins & Porpoises
pp. 49–55

Seals
pp. 56–58

Hoofed Mammals
pp. 59–60

Cats
pp. 60–61

Dogs
pp. 62–63

Bears
p. 64

Weasels & Skunks
pp. 65–67

Raccoon
p. 67

Hares
p. 68

Beaver
p. 68

Porcupine
p. 69

Squirrels
pp. 69–70

Mice, Voles & Kin
pp. 70–72

Moles & Shrews
p. 73

Bats
pp. 73–75

Blue Whale

Balaenoptera musculus

Length: up to 30 m; average 20–27 m
Weight: up to 136,000 kg; average 108,000 kg

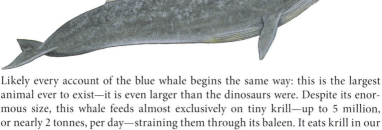

Likely every account of the blue whale begins the same way: this is the largest animal ever to exist—it is even larger than the dinosaurs were. Despite its enormous size, this whale feeds almost exclusively on tiny krill—up to 5 million, or nearly 2 tonnes, per day—straining them through its baleen. It eats krill in our polar waters and migrates south to breed and calve. • This whale is critically endangered. **Where found:** offshore but sometimes in shallow inshore waters; from the Arctic Circle to Panama, including the northwestern Gulf of Mexico. **Also known as:** sulphur-bottom whale (historically, because its belly was often coloured yellow from diatoms accumulated in cold waters), great northern rorqual.

Fin Whale

Balaenoptera physalus

Length: up to 26 m; average 18–21 m
Weight: 27,200–72,575 kg

When this long, sleek giant swims leisurely and gracefully along the surface of the water, its tall, narrow, dense blow reaches over 6 m in height and is very noticeable on the horizon, but the whale does not show its flukes when beginning a dive. It is named for its recognizable and easily seen crescent-shaped dorsal fin. • The fin whale is found singly or in pairs but more often occurs in pods of 3 to 7 individuals, and on occasion several pods have been observed in a small area, creating concentrations of as many as 50 animals. • This whale is an exceptionally fast mover—it has been clocked at speeds over 32 km/hr in short bursts. It is a deep diver and is capable of breaching clear out of the water. **Where found:** inshore and offshore waters; from the Arctic Circle to the Greater Antilles, including the Gulf of Mexico.

Humpback Whale

Magaptera novaeangliae

Length: up to 16 m; average 11.5–15 m
Weight: up to 48,000 kg; average 20,800–27,200 kg

The haunting songs of the humpback last from a few minutes to a few hours and can endure as epic days-long concerts. They have inspired both scientists and artists and reach out to the imaginations of the many people who listen and wonder what this great creature is saying. • This rorqual employs a unique hunting strategy—it creates a bubble net to round up its prey into a tight cluster that the whale then ingests in a food-dense gulpful. **Where found:** along the continental shelf or island banks, sometimes in open offshore waters; from northern Iceland and western Greenland south to the West Indies, including the northern and eastern Gulf of Mexico.

Atlantic Minke Whale

Balaenoptera acutorostrata

Length: up to 10 m; average 8 m
Weight: up to 13,600 kg; average 5000–10,000 kg

The smallest of the rorquals, the minke whale is occasionally seen in our waters, but its seasonal distribution is governed by food availability. • The minke has been one of the more heavily hunted of the baleen whales since the 1980s, when populations of larger whale species had already collapsed. **Where found:** open offshore waters, sometimes in bays, inlets and estuaries; migrates seasonally between warm and cold waters; from the Arctic to the Lesser Antilles, including the eastern and northwestern Gulf of Mexico. **Also known as:** piked whale, sharp-headed finner, little finner, lesser finback, lesser rorqual.

Sei Whale

Balaenoptera borealis

Length: up to 19 m; average 12–16 m
Weight: 14,000–18,000 kg

Populations of sei whales were severely depleted by over-hunting in the 1960s and '70s. Despite their overall small numbers, these whales can be locally abundant in "sei whale years." Typically seen singly or in small groups, they can occur in groups of up to 30 in areas with abundant food. • Although sei whales don't inhabit northern waters, they favour Subarctic feeding grounds in summer and migrate to warmer waters in winter. They eat fish, squid and crustaceans such as krill. **Where found:** throughout offshore temperate waters ranging as far north as the southern tip of Greenland. **Also known as:** sardine whale, pollack whale, coalfish whale, Japan finner, Rudolphi's rorqual.

North Atlantic Right Whale

Eubalaena glacialis

Length: up to 18 m; average 10–16 m
Weight: up to 106,000 kg; average 27,200–72,500 kg

For early whalers, this species was the "right" whale to hunt because it swam slowly and did not sink when it was dead, and the name stuck. The whale was also valued because it yielded large amounts of oil, for fuel, and baleen, for corsets and other uses. The North Atlantic right whale was so heavily hunted that we nearly lost this magnificent creature. This critically endangered species is fully protected from hunting, but populations may never recover. **Where found:** shallow nearshore waters and large bays, as well as offshore; from Iceland to eastern Florida, occasionally into the southern Gulf of Mexico.

Long-finned Pilot Whale

Globicephala melas

Length: 3.8–6 m
Weight: 1600–3200 kg

Acrobatic pilot whales give great performances of some of the most amusing physical behaviours seen in whales: they spyhop (raise their heads straight up above the water's surface to take a look around), lobtail (raise their tail flukes above the water's surface) and slap their flukes on the water. All these great gestures may be a way of attracting attention in lieu of actually shouting out, "Hey! Look over here!" We can only speculate about why whales do the things they do, but they are fun to watch. Another thing we don't yet understand is why whales strand themselves on beaches, and this whale is frequently a victim to this phenomenon. • Squid are the favourite prey of these whales. **Where found:** offshore waters and bays, sometimes inshore in summer; from Iceland and Greenland south to North Carolina.

Beluga

Delphinapterus leucas

Length: up to 5 m; average 3.5 m
Weight: up to 1300 kg

The beluga is a toothed whale of Arctic and Subarctic waters with an isolated population in the St. Lawrence River, one of the most intensively used waterways in Canada. Belugas in the St. Lawrence are subject to all sorts of pollutants, which have negatively affected the health and long-term survival of this population. • A baby beluga is born dark grey, gradually lightening to pure white upon reaching adulthood at 5 to 10 years of age. • This whale is called the "canary of the sea" because of its repertoire of singing, whistling and clicking sounds used for communication and echolocation. **Where found:** St. Lawrence and Saguenay rivers, northeastern and southern Gulf of St. Lawrence, around Newfoundland and NS and along the Labrador coast.

Orca

Orcinus orca

Length: up to 9.8 m; average 5.5–8.5 m
Weight: up to 10,000 kg; average 6800 kg

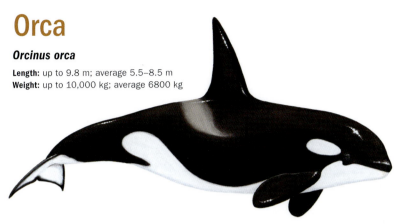

Few people would not recognize these iconic creatures. Found around the world, orcas are revered in First Nations culture and legends, celebrated by enthusiastic whale watchers and, unfortunately, cheered on for entertainment in captivity in aquariums. Referred to as "wolves of the sea," they are strategic group hunters, feeding on fish, squid, sea turtles, sea birds, seals and other whales. • The male orca is larger than the female and has a tall, erect dorsal fin. Females and juveniles have crescent-shaped dorsal fins. **Where found:** cooler coastal waters, offshore waters and bays; from the Arctic to the Lesser Antilles, including the Gulf of Mexico. **Also known as:** killer whale.

Risso's Dolphin

Grampus griseus

Length: up to 4 m; average 2.6–3 m
Weight: up to 500 kg; average 400 kg

Risso's dolphins have the interesting social behaviour of scratching and biting each other, leaving white scars all over their bodies—some older individuals are so scarred that they appear almost completely white. These dolphins can also become scarred from being stung by large squid, their preferred prey. • Although they are typically observed in groups of about a dozen, Risso's dolphins can occur in groups of several hundred, and they become quite engaged in play sessions of breaching, spyhopping, lobtailing and flipper and fluke slapping. **Where found:** deep offshore waters; from eastern Newfoundland to the Lesser Antilles, including the northern and eastern Gulf of Mexico.

Bottlenose Dolphin

Tursiops truncatus

Length: up to 4 m; average 2–3 m
Weight: up to 650 kg; average 200 kg

Most familiar to the public through aquariums, TV shows and various celebrity endorsements, the bottlenose dolphin is intelligent and is one of the most highly studied cetaceans by marine biologists and behavioural ecologists. Individual bottlenose dolphins around the world are being identified and catalogued by markings on their skin and dorsal fins, which are as distinctive and personal as fingerprints. • This dolphin uses echolocation to find prey and communicates primarily vocally. It is playful and acrobatic and is a pleasure to watch surfing the waves or bow riding in front of boats and ships. **Where found:** inshore waters, bays and estuaries all along the coast and even freshwater rivers; from NS to Venezuela, including the Gulf of Mexico.

Short-beaked Common Dolphin

Delphinus delphis

Length: up to 2.6 m; average 1.7–2 m
Weight: up to 140 kg; average 80 kg

Brilliant acrobatic feats accompany the thrill of having a group of these dolphins swim alongside your boat. They love to bow ride and can occur in very large groups of 50 up to 1000 individuals. • A dolphin has the ability to rest one half of its brain at a time, allowing for a constant state of awareness. • The long-beaked common dolphin (*D. capensis*) is very similar both physically and behaviourally to the short-beaked, but does not range in our waters. **Where found:** offshore along the coast; from Newfoundland and NS to northern South America.

Atlantic White-sided Dolphin

Lagenorhynchus acutus

Length: 1.9–2.7 m; female larger than male
Weight: 165–200 kg

Whale watchers lucky enough to see Atlantic white-sided dolphins often get some bonus entertainment from these highly acrobatic animals, which breach, somersault, bow ride and seemingly get very excited at any opportunity to show off. Unlike humans, dolphins can focus their vision above and below water and often take a closer look at people by jumping alongside a boat or lifting their heads above the water's surface. **Where found:** open ocean; increased observations in coastal and sheltered waters, especially between islands and the mainland; from Greenland to New England. **Also known as:** lag, Atlantic striped dolphin, white-striped dolphin, hook-finned dolphin.

Harbour Porpoise

Phocoena phocoena

Length: 1.4–1.9 m
Weight: 55–65 kg

Although commonly seen because of its preference for inshore waters, the harbour porpoise is wary of boats and will not swim alongside them or bow ride. It will instead swim quietly along the surface of the water, feeding on octopus, squid and fish such as herring and doing its best to avoid large sharks and orcas, its main predators. **Where found:** Subarctic and cold-temperate inshore waters, in bays, harbours, estuaries and even at the mouths of rivers; in the Davis Strait and from southeastern Greenland to North Carolina.

Harbour Seal

Phoca vitulina

Length: 1.2–1.7 m
Weight: 50–140 kg

Year-round, large colonies of harbour seals can be observed either basking in the daytime or sleeping at night on rocky shores and islands. Oftentimes during the day, individuals can be seen bobbing vertically in the water. These seals are shy of humans but do occasionally pop their heads up beside a canoe or kayak to investigate, making a quick retreat thereafter. When they disappear below the surface, they are able dive to depths of over 90 m—a feat accomplished by going without breathing for up to 30 minutes. **Where found:** bays, estuaries, intertidal sandbars, rocky shorelines and mouths of rivers along the coast throughout Atlantic Canada; from Baffin I. and Hudson Bay coasts south to the Carolinas.

Ringed Seal

Phoca hispida

Length: 1–1.5 m; male is usually longer than female
Weight: 50–70 kg

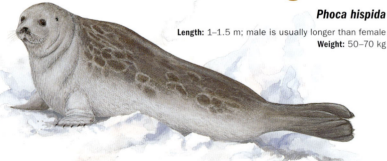

These seals live on the ice—shorefast or drifting pack ice—for most of the year. When the ice starts to freeze in late autumn, the seals create breathing holes that can be up to 2 m long to reach the water below. • Once the winter snow begins to drift and pile, these seals begin digging extensive, multi-chambered lairs. An individual seal will maintain 1 or 2 lairs, which can be up to 4.5 km apart. The white pups are born in these sheltered dens in early spring, where they are protected from the cold and predation by polar bears. If a ringed seal successfully avoids polar bears and other predators, it can live up to 43 years. • Adults have a dark silver pelt marked with dark rings that give this species its name. **Where found:** ice-inhabiting; throughout the Arctic Ocean, ranging as far south as NL. **Also known as:** Arctic ringed seal.

Grey Seal

Halichoerus grypus

Length: *Male:* up to 2.5 m; *Female:* rarely over 2 m
Weight: *Male:* 170–340 kg; *Female:* 100–200 kg

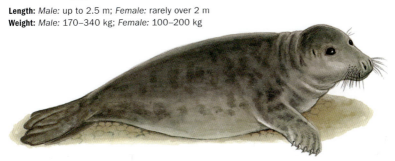

Gregarious grey seals haul out onto the shores of rocky coasts in large groups to breed and moult. After the winter breeding season, these seals disperse widely to feed in pelagic waters. Even pups, soon after being weaned, have been tracked 1000 km away from the shores of their birth. • These seals dive 30 to 70 m to feed on fish, crustaceans and cephalopods, and have been recorded diving to depths of 300 m and remaining submerged for up to 20 minutes. • There are 2 grey seal herds in Atlantic Canada, which breed primarily in the southern Gulf of St. Lawrence and on Sable Island, Nova Scotia. **Where found:** rocky shores throughout Atlantic Canada; Gulf of St. Lawrence and from Labrador to New England. **Also known as:** horsehead seal.

Harp Seal

Phoca groenlandica

Length: 1.4–2 m; average 1.7 m
Weight: average 130 kg

This high-endurance seal embarks upon annual migrations of up to 9650 km and can to dive to 275 m, holding its breath for up to 15 minutes. • Pups are born on the edge of the ice pack in February and are covered in white fur called lanugo. After nursing for 2 weeks, the pup grows from its birth weight of 5.5 kg to up to 45 kg, at which time the fat pup is left to fend for itself. Over the next couple of weeks, it drops half of its weight and moults its downy lanugo. Those that do not starve to death or fall prey to human hunters, polar bears or orcas eventually learn to fish and migrate north with the herd in summer. **Where found:** in winter on drifting pack ice in offshore waters; throughout Atlantic Canada, occasionally up streams; breeds in the Gulf of St. Lawrence and off the coast of NL. **Also known as:** *Pagophilus groenlandicus.*

Bearded Seal

Erignathus barbatus

Length: 2–2.5 m
Weight: 200–250 kg; maximum 360 kg

The bearded seal's face, with its long whiskers and small head, makes this creature easily distinguishable. The coat is unpatterned, ranging in colour from grey to brown. Pups lose their white fur (lanugo) while still in the womb and are born with a greyish brown coat. • Bearded seals have a varied diet that consists mainly of crustaceans such as shrimp and crabs and molluscs such as clams and whelks. Some fish species, such as sculpin, flatfish and cod, are also eaten. These seals prefer to feed at the ocean floor in areas with water depths of less than 130 m. **Where found:** prefers seasonally ice-covered waters less than 200 m deep, areas of broken pack ice and drifting ice floes; mainly Arctic ranging but also inhabits the North Atlantic Ocean around NL, occurring as far south as the Gulf of St. Lawrence.

Hooded Seal

Cystophora cristata

Length: 1.8–3 m; average
male 2.5 m; average female 2 m
Weight: *Male:* 300–400 kg;
Female: 160–230 kg

The hood of this seal is a bulbous, red mass on the male's head. The bull inflates the hood to make himself look more formidable to any aggressors that anger or threaten him. • Breeding, calving and pup moulting occurs throughout the winter on the drifting pack ice. This seal is abundant in winter off the Grand Banks of Newfoundland, where there are large schools of fish to feed on so the seals can fatten up to survive the long cold season. In summer, the species disperses throughout the North Atlantic Ocean. • This seal can dive to depths of over 1000 m and stay submerged for over 50 minutes. Individuals can live for 30 to 35 years. **Where found:** on the edge of drifting pack ice, in deep offshore waters of the North Atlantic; Labrador, northeastern Newfoundland and the Gulf of St. Lawrence.

Moose

Alces alces

Length: 2.5–3 m (including 9–19 cm tail)
Shoulder height: 1.7–2.1 m
Weight: 230–600 kg

The largest deer in the world, moose have been known to dive to depths of 4 m to find aquatic plants rich in salts and minerals—and to escape those nasty biting insects! • Moose browse on trees and shrubs, and graze on grasses and forbs. • Bulls have large, wide antlers that can measure up to 1.5 m across. A distinguishing dewlap, or "bell," a long flap of fur-covered skin, hangs from the throat. **Where found:** near lakes and bogs; in riparian valleys, coniferous forests and willow and poplar groves; Newfoundland (introduced) and Labrador, the QC north shore and north of the Gulf of St. Lawrence.

White-tailed Deer

Odocoileus virginianus

Length: 1.4–2.1 m
(including 21–36 cm tail)
Shoulder height: 70–115 cm
Weight: 50–200 kg

A wagging white tail disappearing into the forest is a common view of this deer. • When a mother deer is feeding, it leaves its scentless, spotted fawn behind among tall grasses or shrubs to hide it from potential predators. • A dense network of blood vessels covered by hair, called "velvet," covers the developing antlers of males in spring and summer. **Where found:** rolling country with open areas near cover; valleys and stream courses, woodlands, meadows and abandoned farmsteads with tangled shelterbelts; absent only from NL and the extreme north shore of QC. **Also known as:** Virginia deer.

Woodland Caribou

Rangifer tarandus ssp. *caribou*

Length: 1.4–2.1 m (including 21–36 cm tail)
Shoulder height: 70–115 cm
Weight: 50–200 kg

The woodland caribou is the only member of the deer family in North America with both sexes having antlers; the female's are slender, and the male's are large and C-shaped. Large, crescent-shaped hooves, a long throat mane and well-furred ears and muzzle make the caribou superbly adapted for surviving harsh winters. • Caribou migrate to different elevations between seasons. In winter, tree lichens form the bulk of their diet. • The Atlantic-Gaspésie population southeast of the St. Lawrence River is a relic herd of the last remaining maritime woodland caribou. **Where found:** old-growth coniferous forests, tundra and subalpine and alpine meadows; NL and the QC north shore.

Mountain Lion

Puma concolor

Length: 1.5–2.7 m (including 50–90 cm tail)
Shoulder height: 65–80 cm
Weight: 30–90 kg

This secretive cat is seldom seen by people, but occasional track or scratch marks and rare sightings have been reported in New Brunswick. • The mountain lion prefers to sit in a tree above an animal trail and pounce on its prey, which is mainly deer, but opportunities to take other prey—other ungulates, beavers, rabbits or birds—will not be passed up. **Where found:** montane regions; may occur in brushlands or subalpine regions based on food availability; long extirpated from the East, yet reports abound in QC, NB and NS (over 1000 sightings have been reported in NS and NB since 1949). **Also known as:** cougar, puma; *Felis concolor*.

Lynx

Lynx canadensis

Length: 80–100 cm
(including 9–12 cm tail)
Shoulder height: 45–60 cm
Weight: 7–18 kg

With long legs and huge, well-furred paws, the lynx is uniquely adapted for catching snowshoe hares on snow. There is a very close predator-prey relationship between the lynx and the hare. Cyclical increases and decreases in hare populations, which are governed largely by food availability, cause lynx populations to follow similar trends—when hares are abundant, lynx kittens are more likely to survive and reproduce. • This feline's facial ruff, long, black ear tufts and short, black-tipped tail are distinctive features. The coat is grey to orange-brown. **Where found:** boreal, dense, old-growth coniferous forests with heavy undergrowth; St. Lawrence region, Gaspé Peninsula, QC north shore, Cape Breton I., NB and NL.

Bobcat

Lynx rufus

Length: 75–125 cm
(including 13–17 cm tail)
Shoulder height: 45–55 cm
Weight: 7–13 kg

The nocturnal bobcat feeds on a wide range of prey, including rabbits, voles, mice, birds, reptiles and insects. Small but mighty, the bobcat is even capable of bringing down a deer by the throat if the opportunity presents itself. • This cat's atypically short, "bobbed" tail is well suited to the shrubby and forested areas in which it hunts, but the bobcat is highly adaptable and may even be seen close to residential areas. • Like most young cats, bobcat kittens are almost always at play. **Where found:** coniferous and deciduous forests and brushy areas, riparian areas of willow stands; Gaspé Peninsula and throughout the Maritimes.

Grey Wolf

Canis lupus

Length: 1.35–2 m
(including 35–50 cm tail)
Shoulder height: 60–90 cm
Weight: 25–36 kg

The hauntingly beautiful howl of the grey wolf is a unique and vital wilderness sound. However, this animal was not always appreciated and was exterminated as vermin from much of its range. • Wolf packs have a strong social hierarchy, with often only one breeding pair in each pack. • The wolf has grey, white or occasionally black pelage. It has a thicker, wider muzzle than a coyote and holds its tail high when running. **Where found:** boreal forest, tundra and remote wilderness areas; Labrador and QC, extirpated from the islands. **Also known as:** timber wolf, tundra wolf, Arctic wolf, Labrador wolf.

Coyote

Canis latrans

Length: 1.2–1.5 m (including 30–40 cm tail)
Shoulder height: 58–66 cm
Weight: 9–23 kg

Occasionally forming loose packs and joining in spirited yipping choruses, coyotes are intelligent and versatile hunter-scavengers, best described as opportunistic omnivores. They have been observed fishing or even engaging the help of a hunting badger to catch ground squirrels. • The size of an average dog, coyotes share many characteristics that we appreciate in domestic canines, but they are companions only to the wilderness and their fellow pack members. **Where found:** mixed and coniferous forests, meadows, agricultural lands and suburban areas; southern QC, NB, NS and Newfoundland. **Also known as:** brush wolf.

Red Fox

Vulpes vulpes

Length: 90–110 cm (including 35–45 cm tail)
Shoulder height: 38–45 cm
Weight: 3.6–6.8 kg

The red fox is a talented and entertaining mouser with high-pouncing antics that are much more cat-like than canine. Its extroverted behaviour and noble good looks have landed it roles in many fairy tales, fables and Native legends. • This fox is typically a vivid reddish orange but can have darker colour phases, with dark fur across the back and shoulders, or the coat can be almost entirely black with silver-tipped hairs. The tip of its elegant, bushy tail, however, is always white. **Where found:** prefers open, grassy habitats with brushy shelter, riparian areas and forest edges but avoids dense forests; throughout Atlantic Canada. **Also known as:** silver fox (referring to a colour variation).

Arctic Fox

Vulpes lagopus

Length: 85 cm (including 31 cm tail)
Shoulder height: 25–30 cm
Weight: *Male:* 3.5 kg; *Female:* 3 kg

The Arctic fox's distinguishing feature is its beautiful, thick, white winter pelage, but in summer, the coat thins out and darkens to a bluish brown. • Found in family units only during breeding and pup rearing, this fox does not seek the company of its own kind but is more commonly seen in winter following polar bears to scavenge on their prey. In summer, this fox is much more self-reliant and hunts numerous small prey species. **Where found:** typically above treeline in tundra, Subarctic and Arctic regions; northern QC and Labrador. **Also known as:** *Alopex lagopus.*

Black Bear

Ursus americanus

Length: *Male:* 1.3–1.9 m;
Female: 1.2–1.6 m
Shoulder height: 60–110 cm
Weight: *Male:* 120–400 kg;
Female: 40–230 kg

The black bear is primarily a forest dweller, with long claws well adapted to climbing trees and digging. It is omnivorous, eating plants, insects such as bees (and honey) and carrion. It sometimes even preys upon small rodents or young deer. This bear is prone to becoming habituated to humans by finding food in compost piles, garbage bins, granaries, bird feeders and the like. • Although variable in pelage in the West, eastern black bears almost always have the typical glossy, black fur with a tan muzzle, and many have a white "V" on the chest. **Where found:** mixed forests and shrub thickets with nut- and berry-producing plants; also swamps and suburban areas; absent only from PEI.

Polar Bear

Ursus maritimus

Length: *Male:* 2–3 m; *Female:* 2–2.5 m
Shoulder height: up to 1.6 m
Weight: *Male:* 300–800 kg;
Female: 150–300 kg (doubling
when pregnant)

The unmistakable great, white polar bear is the largest bear and the largest quadruped carnivore on the planet. A large male, when standing on his hind legs, may approach 3 m in height to the top of his head. Of the world's polar bears (roughly estimated at 21,000 to 28,000 individuals), 60 percent (or approximately 15,000) occur in or are shared with Canada's eastern Arctic. • Although the polar bear is terrestrial, it is well adapted to swimming in the frigid Arctic Ocean and is an integral member of the marine ecosystem. It feeds primarily on seals. The diminishing Arctic ice is a concerning case of habitat loss for this iconic species. **Where found:** on land, landfast ice and permanent offshore pack ice; along the coastline of the polar basin and Arctic islands; northern QC, NL and the Gulf of St. Lawrence (50° N latitude) mark its most southerly range.

Wolverine

Gulo gulo

Length: 70–110 cm (including 17–25 cm tail)
Weight: 7–16 kg

This large member of the weasel family looks like a small, frazzled bear. The wolverine has a bad reputation because it occasionally raids unoccupied wilderness cabins, consuming anything edible and spraying any leftovers with foul-smelling musk from its anal glands. Its wild prey includes marmots, ground squirrels, gophers, mice, insects and berries; it will also scavenge carrion. **Where found:** remote, wooded foothills and mountains; alpine tundra in summer, lower elevations in winter; Labrador and northern QC (critically endangered so any sightings should be reported). **Also known as:** glutton, skunk bear.

Marten

Martes americana

Length: 50–68 cm (including 18–23 cm tail)
Weight: 0.5–1.2 kg

An expert climber with semi-retractable claws, this forest dweller is agile enough to catch arboreal squirrels. Although it spends most of its time on the ground in search of rodent prey, the marten often dens in a tree hollow, where it raises its annual litter of 1 to 5 kits. • This elusive animal sometimes falls victim to traplines, an ongoing threat even today. **Where found:** old-growth boreal and montane coniferous forests with numerous dead trunks, branches and leaf cover; QC north shore, Gaspé Peninsula, St. Lawrence region, NB, Cape Breton I. and Newfoundland. **Also known as:** American sable, pine marten.

Fisher

Martes pennanti

Length: 80–120 cm (including 30–40 cm tail)
Weight: 2–5.5 kg

Despite the name, fishers rarely consume fish but prey upon rodents, hares and birds, as well as eating berries, nuts and sometimes carrion. They will eat any animal they can overpower but are distinguished, along with mountain lions, for their ability to prey upon porcupines. • Fishers are extremely sensitive to any human disturbance and exist only in remote, forested wilderness. **Where found:** dense, mixed and coniferous forests (absent from young, thinly treed, logged or burned forests); southern QC, Gaspé Peninsula, St. Lawrence region, NB and NS.

Long-tailed Weasel

Mustela frenata

Length: 28–42 cm (including 12–29 cm tail)
Weight: 85–400 g

Following the tracks of the long-tailed weasel on a snow-covered meadow offers good insight into the curious and energetic nature of this little mammal. Constantly distracted from walking in a straight line, it continuously zigs and zags to investigate everything that catches its attention. • This weasel feeds on small rodents, birds, insects, reptiles, amphibians and occasionally fruits and berries. • Like other true weasels, it turns white in winter, but the tip of the tail remains black. **Where found:** aspen parklands, intermontane valleys and open forests; throughout Atlantic Canada.

Short-tailed Weasel

Mustela erminea

Length: 20–35 cm (including 4–9 cm tail)
Weight: 45–105 g

The short-tailed weasel is a voracious nocturnal hunter of mice and voles. Although relatively common, it will not linger for any admiring observers; a spontaneous encounter with this curious creature will reveal its extraordinary speed and agility as it quickly escapes from view. • This weasel's coat is white in winter, but the tail is black-tipped year-round. **Where found:** coniferous and mixed forests; in summer, often found in alpine tundra, where it hunts on rock slides and talus slopes; throughout QC, Labrador, NB and NS. **Also known as:** ermine, stoat.

Mink

Neovison vison

Length: 47–70 cm (including 15–20 cm tail)
Weight: 0.6–1.4 kg

The mink's partially webbed feet make it an excellent swimmer, and it is capable of diving to depths of more than 3 m in pursuit of fish. Its thick, dark brown to blackish, oily fur insulates its body from extremely cold water. • The mink travels along established hunting routes, often along shorelines, rarely foregoing a prey opportunity. It stashes any surplus kills in temporary dens, typically dug into riverbanks, beneath rock piles or in evacuated muskrat lodges. **Where found:** shorelines of lakes, marshes and streams in forests and woods in foothills and on grasslands; throughout Atlantic Canada. **Also known as:** *Mustela vison*.

Northern River Otter

Lontra canadensis

Length: 0.9–1.4 m (including 30–50 cm tail)
Weight: 5–11 kg

The favourite sport of these frisky otters is sliding down riverbanks, wet, grassy hills and even snowy slopes in winter—look for their "slides" on the banks of rivers, lakes and ponds. When otters are not at play, they are engaged in the business of hunting. These swift swimmers mainly prey upon aquatic species such as crustaceans, turtles, frogs and fish, but they occasionally depredate bird nests and eat small rodents. **Where found:** fresh- and saltwater habitats; lakes, ponds and streams; also along the coast; throughout Atlantic Canada. **Also known as:** *Lutra canadensis.*

Striped Skunk

Mephitis mephitis

Length: 55–80 cm (including 20–35 cm tail)
Weight: 1.9–4.2 kg

Butylmercaptan is responsible for the stink of the striped skunk's musk, which is sprayed in self-defence. Only the great horned owl is undeterred by the skunk's odour and is one of this mammal's few predators. When undisturbed, the striped skunk is a quiet, reclusive omnivore, feeding on insects, worms, bird eggs, reptiles, amphibians, grains, green vegetation, berries and, rarely, small mammals and carrion. **Where found:** lower-elevation streamside woodlands, groves of hardwood trees, semi-open areas, brushy grasslands and valleys; also urban areas; throughout the Maritimes and southern QC.

Raccoon

Procyon lotor

Length: 65–100 cm (including 19–40 cm tail)
Weight: 5–14 kg

Garbage containers are no match for the raccoon's curiosity, persistence and problem-solving abilities, making them and garden goldfish ponds prime targets for midnight food raids in urban areas. In this animal's natural habitat, an omnivorous diet of clams, frogs, fish, bird eggs and nestlings, berries, nuts and insects is more than ample. • The raccoon builds up its fat reserves during the warm months to sustain itself through the winter. **Where found:** lower-elevation riparian areas or edge habitats between forests and wetlands such as streams, lakes and ponds; throughout the Gaspé Peninsula and the Maritimes.

Arctic Hare

Lepus arcticus

Length: 55–70 cm (including 40–70 mm tail)
Weight: 3–7 kg

When the Arctic hare sports its brilliant white winter coat, the tips of its ears remain black. Its ears are shorter than those of any other hare, an adaptation for conserving heat in winter. Come summer, this hare's pelage turns blue-grey or grey-brown to blend in with the austere colours of the bare tundra. • Snowy owls and other raptors, weasels, foxes and polar bears prey upon the Arctic hare, but its main predator is the wolf, which it can sometimes evade by running at speeds up to 60 km/hr. • A litter of typically 5 young, called leverets, are born in May. **Where found:** tundra and lowland areas where there is sufficient vegetation; NL. **Also known as:** polar hare, polar rabbit.

Snowshoe Hare

Lepus americanus

Length: 38–53 cm (including 4.8–5.4 cm tail)
Weight: 1–1.5 kg

Extremely well adapted for surviving harsh alpine winters, the snowshoe hare has large hind feet that allow it to move across deep snow without sinking, while the white pelage camouflages the animal. If detected by a predator, the hare explodes into a running zigzag pattern, reaching speeds of up to 50 km/hr. • Populations of this hare, its winter food sources of willow and alder and its main predator, the lynx, are closely interrelated. **Where found:** brushy, second-growth forests, boreal and hardwood forests; introduced but possibly absent from Newfoundland, otherwise throughout Atlantic Canada. **Also known as:** varying hare.

Beaver

Castor canadensis

Length: 90–120 cm (including 28–53 cm tail)
Weight: 16–30 kg

The loud slap of a beaver's tail on water warns of intruders, and the tail is also an extremely effective propulsion device for swimming and diving. • The beaver's long, continuously growing incisors help the animal cut down trees, and its strong jaws allow it to drag pieces of wood weighing up to 9 kg. • Beaver was the most sought-after pelt during the fur trade, with one beaver pelt forming a unit of currency against which all other pelts were evaluated. **Where found:** lakes, ponds, marshes and slow-flowing rivers and streams with ample vegetation; throughout Atlantic Canada.

Porcupine

Erethizon dorsatum

Length: 55–95 cm (including 14–22 cm tail)
Weight: 3.5–18 kg

Porcupines do not throw their 30,000 or so quills but deliver them into the flesh of an attacker with a quick flick of the tail. • This excellent tree climber fills its vegetarian diet with forbs, shrubs and the sugary cambium of trees. An insatiable craving for salt occasionally drives it to gnaw on rubber tires, wooden axe handles, toilet seats and even hiking boots! **Where found:** coniferous and mixed forests, open tundra and even rangelands; absent from Newfoundland and some islands, otherwise throughout Atlantic Canada.

Woodchuck

Marmota monax

Length: 46–66 cm (including 11–16 cm tail)
Weight: 1.8–5.4 kg

Also known as "groundhog," this animal is rarely even thought about until February 2, when everyone anticipates the sleepy creature emerging from its den to see its shadow. However, you are unlikely to see a woodchuck above ground before April or May. • The woodchuck's den is typically several metres deep, and abandoned dens are taken over by many other species, such as foxes. **Where found:** in or alongside their burrows in rock piles, ravines, open woodlands, pastures, meadows, under barns and in natural areas within city limits; throughout Atlantic Canada but absent from Newfoundland and some islands. **Also known as:** groundhog, marmot.

Red Squirrel

Tamiasciurus hudsonicus

Length: 28–35 cm (including 11–15 cm tail)
Weight: 170–310 g

The red squirrel is a common visitor to backyards and town parks, generally causing a racket with its ongoing nattering and chattering—monologues directed at any passerby, either human or animal. It even mutters to itself while busily collecting food for its cache. • An adventurous diner, this squirrel will eat or store any available source of nutrition—pine cones, nuts, seeds, fungi, fruits and also animal protein such as eggs, nestling birds, baby mammals and carrion. The large caches keep the squirrel fed throughout the year, including winter, when it remains active. **Where found:** boreal coniferous and mixed forests; throughout Atlantic Canada. **Also known as:** chickaree.

Northern Flying Squirrel

Glaucomys sabrinus

Length: 25–38 cm (including 11–18 cm tail)
Weight: 75–180 g

Long flaps of skin (called the patagium) stretched between the fore and hind limbs and a broad, flattened tail allow this nocturnal squirrel to glide swiftly from tree to tree, with extreme glides covering distances up to 100 metres! • Flying squirrels play an important role in forest ecology because they dig up and eat truffles, spreading around the fruiting bodies of these beneficial ectomycorrhizal fungi. **Where found:** primarily old-growth, coniferous forests but also aspen and cottonwood woodlands; absent from Newfoundland and some islands, otherwise throughout Atlantic Canada.

Eastern Chipmunk

Tamias striatus

Length: 23–30 cm (including 7–10 cm tail)
Weight: 66–139 g

This common chipmunk of eastern Canada is fond of both city and country living—a natural inhabitant of the forest, it also makes itself at home in backyards and city parks. • The eastern chipmunk becomes quite cute and chubby after a summer of feasting on nuts, seeds, berries, fungi, insects, slugs and snails despite its active lifestyle. It needs this fat reserve to sustain itself throughout the winter, when it hibernates in its underground burrow. **Where found:** open, deciduous woodlands, forest edges, brushy areas, rocky outcroppings and treed urban areas; southern QC and the Maritimes.

Meadow Vole

Microtus pennsylvanicus

Length: 13–19 cm (including 3–5 cm tail)
Weight: 30–64 g

There are several vole species in Atlantic Canada. These little mammals fulfill an important role as a prey species for many predators, including birds of prey, snakes and carnivorous mammals. The meadow vole is likely one of the most abundant species, and it ranges across the country. • This vole is also active in winter, just below the snow. This subnivean habitat is insulated from the elements, contains insects and dried vegetation for food and conceals the vole from predators unless, like the fox, they have particularly good hearing and pounce through the snow layer to catch a winter's meal. **Where found:** grasslands, pastures, marshy areas, open woodlands and tundra; throughout Atlantic Canada.

Southern Bog Lemming

Synaptomys cooperi

Length: 12–15 cm (including 1.3–2.4 cm tail)
Weight: 21–50 g

The southern bog lemming is actually a species of vole, and where its range overlaps that of the meadow vole, the lemming is outcompeted. When it finds sufficient habitat, it occupies itself by clearing paths through the grass, leaving neat piles of clippings—and plenty of droppings—along the route. Its strong, curved claws aid in digging elaborate winter runways and underground burrows, but the lemming may also use the abandoned burrows of other small mammals or build grassy aboveground nests. • Lemmings feed on grasses, sedges and clover, as well as fungi, mosses, roots and even algae. **Where found:** grassy meadows, shrub edges and open forests; from southeastern MB east to Newfoundland.

Common Muskrat

Ondatra zibethicus

Length: 46–61 cm (including 20–28 cm tail)
Weight: 0.8–1.6 kg

Although they share similar habitats and behaviours, the beaver and the common muskrat are not closely related. The muskrat also sports large incisors, which it uses to cut through a vast array of thick vegetation, particularly cattails and bulrushes. It makes a partially submerged den similar to that of a beaver, which provides a nesting spot for many geese and ducks, as well as important shelter for other rodents when the muskrat moves house. **Where found:** low-elevation sloughs, lakes, marshes and streams with plenty of cattails, rushes and open water; throughout Atlantic Canada.

Norway Rat

Rattus norvegicus

Length: 33–46 cm (including 12–22 cm tail)
Weight: 200–480 g

Native to Europe and Asia, the Norway rat came to North America as a stowaway on ships in about 1775. This rodent is mainly associated with human settlements, feeding on cereal grains, fruits, vegetation and garbage, and basically making a nuisance of itself—an example of an introduced species becoming a reviled pest though it survives only in a nonnative environment. • Captive-bred rats have aided scientific research in many fields. **Where found:** urban areas, farmyards, garbage dumps; southern QC and the Maritimes. **Also known as:** brown rat, common rat, sewer rat, water rat.

Deer Mouse

Peromyscus maniculatus

Length: 14–21 cm (including 5–10 cm tail)
Weight: 18–35 g

The abundant deer mouse is a seed eater, but it will also consume insects, spiders, caterpillars, fungi, flowers and berries. It is in turn an important prey species for many other animals, so it must be a prolific breeder to maintain its population. • A litter of 4 to 9 young leaves the nest after 3 to 5 weeks, and the young mice are sexually mature 1 to 2 weeks after that. Less than 5 percent survive a complete year. **Where found:** most dry habitats, grasslands, shrublands, forests and human settings; absent from Newfoundland and some islands, otherwise throughout Atlantic Canada.

House Mouse

Mus musculus

Length: 13–20 cm (including 6–10 cm tail)
Weight: 14–24 g

This familiar mouse can be found throughout most of North America. Like the Norway rat, it arrived as a stowaway on ships from Europe, quickly spreading across the continent alongside early settlers. • The house mouse is nocturnal and may be responsible for gnawing the labels off the canned soup stored in your cupboards! • This mouse's pelage is brownish to blackish grey with grey undersides. **Where found:** usually associated with humans in both rural and urban settings, including houses, garages, farmyards, garbage dumps and granaries; absent from Labrador and northern QC, otherwise throughout Atlantic Canada.

Woodland Jumping Mouse

Napaeozapus insignis

Length: 20.5–25.5 cm (including 11.5–16 cm tail)
Weight: 17–35 g

Named jumping mice for a reason, the woodland species can leap 4 m in a single bound. This agility helps it avoid predators. • This mouse feeds on fruits, seeds, herbs, fungi and roots, which it prefers to forage for at night, as it is nocturnal by nature. **Where found:** along woodland streams in brushy deciduous and coniferous forests; throughout most of the Maritimes, Labrador and QC.

Star-nosed Mole

Condylura cristata

Length: 20 cm (including 8 cm tail)
Weight: 55 g

The 22 finger-like tentacles surrounding this mole's nose, which are tactile and have enough dexterity and strength to manipulate objects, make this rarely seen little animal a star attraction. • The outward-facing front paws with long claws are ideal for digging tunnels through the moist soil in which the star-nosed mole burrows. It is also an excellent swimmer, the paws working like paddles and the tail (which stores fat in the winter) working like a rudder. • The star-nosed mole feeds on insects, worms and aquatic invertebrates. **Where found:** swamps, meadows, marshes, lakes and streambanks; throughout the Maritimes.

Masked Shrew

Sorex cinereus

Length: 10 cm (including 4 cm tail)
Weight: 2–7 g

The masked shrew is likely the most abundant mammal in Atlantic Canada, including offshore islands. It is active year-round, and in winter, it inhabits the subnivean layer beneath the snow but aboveground. • The large populations of small rodents such as shrews and voles sustain the high predation rates these animals suffer; they are important prey species year-round for small carnivores such as foxes and weasels, and birds of prey such as owls. **Where found:** most habitats except very wet ones; throughout Atlantic Canada.

Big Brown Bat

Eptesicus fuscus

Length: 9–14 cm (including 2–6 cm tail)
Wingspan: 33 cm (forearm 4.1–5.4 cm)
Weight: 12–28 g

An effective aerial hunter, the big brown bat uses ultrasonic echolocation (80,000 to 40,000 Hz) to detect flying insects up to almost 5 m away. It flies above water, around street lights and over agricultural areas hunting insects at dusk and dawn. • This bat is not abundant but is frequently encountered because of its tendency to roost in human-made structures. It has been known to change hibernation sites midwinter, a time when it is extremely rare to spot a bat. **Where found:** in and around human-made structures; occasionally roosts in hollow trees and rock crevices; southern QC, the Gaspé Peninsula and NB.

Eastern Pipistrelle

Pipistrellus subflavus

Length: 7–9 cm (including 3.4–4.1 cm tail)
Wingspan: 21–26 cm (forearm 3–3.5 cm)
Weight: 6–10 g

Quite delicate, able to fit in a matchbox and with a weak, erratic flight pattern, this pip of a pipistrelle is unable to fly in a strong wind, yet it is stronger than its western counterpart. Some individuals even migrate several hundred kilometres in late summer and early autumn to the caves where they hibernate for the winter. • Females are larger than males and give birth to twins. Lifespan records for this species are 15 years for males and 10 years for females. **Where found:** in caves in winter; in summer, roosts in trees, buildings and barns, and on cliffs; the northern limit of its range is southern coastal NB and southern NS.

Eastern Red Bat

Lasiurus borealis

Length: 10–11 cm (including 4.5–6.2 cm tail)
Wingspan: 29–33 cm (forearm 3.7–4.5 cm)
Weight: 7–16 g

The insectivorous, forest-dwelling eastern red bat hunts low to the ground, catching flying ants, moths, leaf hoppers and beetles. • Active at night, this bat usually roosts in trees by day, preferring those with dense foliage for concealment, but it will also roost in the open or sometimes in a cave or tunnel. It is quite solitary and does not roost in colonies. • The red fur often has white tips, typically in the female, giving the coat a frosted look. **Where found:** deciduous forests; throughout the Maritimes and the Gaspé Peninsula in summer.

Hoary Bat

Lasiurus cinereus

Length: 11–15 cm (including 4.1–6.7 cm tail)
Wingspan: 40 cm (forearm 4.5–5.7 cm)
Weight: 19–35 g

This large, beautiful bat roosts in trees, not caves or buildings, and wraps its wings around itself for protection against the elements, the frosty-coloured fur blending in among the mosses and lichens. The hoary bat also roosts in orchards, but it is an insectivore and does not damage fruit crops. At night, this bat can be recognized by its large size and slow wingbeats over open terrain. **Where found:** in open areas and around lakes near coniferous and deciduous forests; southern QC, NB and NS.

Little Brown Bat

Myotis lucifugus

Length: 7–10 cm (including 2.5–5.4 cm tail)
Wingspan: 25 cm (forearm 3.5–4.1 cm)
Weight: 5.3–8.9 g

On warm, calm summer nights, the skies are filled
with the shrill calls of bats, but the frequencies are
beyond the range of our hearing. There are several species
of mouse-eared bats (*Myotis* spp.) in Atlantic Canada, but
they are generally indistinguishable from each other as they fly in dim light. The
little brown bat is most commonly seen skimming over lakes, ponds and even
swimming pools, taking a quick drink and pursuing insects. **Where found:** roosts
in buildings, barns, caves, rock crevices, hollow trees and under tree bark; hibernates in buildings, caves and mines; southern QC, the Gaspé Peninsula and NB.
Also known as: little brown myotis.

Silver-haired Bat

Lasionycteris noctivagans

Length: 9–11 cm (including 3.5–5.1 cm tail)
Wingspan: 30 cm (forearm 4.1–5.4 cm)
Weight: 4–12 g

This bat takes flight at twilight, at both dawn and dusk,
embarking on feeding forays for moths and flies over open
fields, water and treetops. • The silver-haired bat prefers to
roost in trees, and to conserve energy on cold days, it can
lower its body temperature and metabolism—a state known
as torpor. • Typically solitary in summer, this bat migrates
southward to warmer climes in winter and forms small colonies that hibernate in caves, mines or abandoned buildings.
Females form nursery colonies in protected shelters such as tree cavities. **Where
found:** roosts in cavities and crevices of old-growth trees but can adapt to parks,
cities and farmlands; throughout southern QC and the Maritimes.

BIRDS

Birds are the most diverse class of vertebrates. All birds are feathered but not all fly. Traits common to all birds are that they are two-legged, warm-blooded and lay hard-shelled eggs. Some migrate south in the colder winter months and return north in spring. For this reason, the diversity of bird species in a region varies with the seasons. Although some species such as the common eider and northern goshawk can be seen year-round, northern climates dictate seasonal migration because few birds can adapt to the extreme changes from hot summer to frozen winter. Some birds change their local address to adapt to the seasons; for example, the common loon can be seen pretty much anywhere in the region when the weather is fine, but winters exclusively on milder coastal ocean waters, which do not freeze in winter.

Wetland birds do not do well when lakes and ponds freeze and therefore most are only seen in summer. Many ducks, herons, sandpipers and rails fall into this category, as well as the osprey, which feeds on fish. Most raptors migrate south for the winter or find the northern extent of their winter range in the southern reaches of the Maritimes. Some shorebirds such as the ruddy turnstone love cooler climes, breeding in the Arctic and heading south for the winter, passing through our region in migration. Some of our well-known winter birds include chickadees, downy woodpeckers, waxwings, nuthatches and snowy owls.

Spring brings scores of migrant waterfowl, colourful songbirds that breed in Atlantic Canada and other birds such as shorebirds that continue on to Arctic breeding grounds. Other migratory birds, such as Canada geese, pass through in autumn, their numbers bolstered by the young of the year. Many species are in duller plumage in fall and winter.

Scores of migrating birds fly as far south as Central and South America. These neotropical migrants are of concern to biologists and conservationists because of habitat degradation and loss, collisions with human-made towers, pesticide use and other factors that threaten their survival. Education and an increasing appreciation for wildlife may encourage solutions to these problems.

Waterfowl
pp. 78–79

Grouse & Allies
p. 79

Diving Birds
pp. 81–82

Herons
pp. 82–83

Birds of Prey
pp. 84–86

Rails
p. 87

Shorebirds
pp. 87–90

Gulls, Terns & Alcids
pp. 90–92

Doves & Cuckoos
p. 93

Owls
pp. 94–95

Nightjars, Hummingbirds
& Kingfishers
p. 96

Woodpeckers
pp. 97–98

Flycatchers
pp. 98–99

Shrikes & Vireos
p. 99

Jays & Ravens
p. 100

Larks & Swallows
p. 101

Chickadees, Nut-
hatches & Wrens
p. 102

Kinglets & Thrushes
pp. 103–104

Mimics, Starlings
& Waxwings
p. 105

Wood-warblers
pp. 106–107

Sparrows & Buntings
pp. 108–109

Blackbirds & Allies
p. 110

Finch-like Birds
pp. 111–112

Canada Goose

Branta canadensis

Length: 55–122 cm
Wingspan: up to 1.8 m

Canada geese mate for life and are devoted parents to their 2 to 11 goslings. • Wild geese can be aggressive when defending their young or competing for food. Hissing sounds and low, outstretched necks are signs that you should give these birds some space. • Geese graze on aquatic grasses and sprouts, and you can spot them tipping up to grab for aquatic roots and tubers. • The Canada goose was split into 2 species in 2004; the smaller subspecies has been renamed the cackling goose (*B. hutchinsii*). **Where found:** near water bodies and in parks, marshes and croplands; throughout NL, PEI and the St. Lawrence region; limited range in NS and NB, where it is more commonly seen in migration.

Wood Duck

Aix sponsa

Length: 38–54 cm
Wingspan: 71–99 cm

A forest-dweller, the wood duck is equipped with sharp claws for perching on branches and nesting in tree cavities, which may be as high as 7.5 to 9 m. Shortly after hatching, the ducklings jump out of their nest cavity, but, like downy ping-pong balls, they bounce upon landing and are seldom injured. • A female wood duck often returns to the same nest site each year; being familiar with potential threats at established nest sites may help her improve her brood's survival rate. **Where found:** swamps, ponds, marshes and lakeshores with wooded edges; summer range in NS, PEI and NB.

American Black Duck

Anas rubripes

Length: 58 cm
Wingspan: 85–95 cm

These common ducks are seen year-round, typically in pairs throughout winter after pair bonds are established in September. They often mingle with mallards (*A. platyrhychos*), with which they frequently interbreed, and the offspring show the colouration and features of both parents. • These ducks eat primarily plant matter throughout summer but feed on invertebrates in winter to obtain heat-generating protein. **Where found:** fresh water, flooded fields, croplands and coastal bays; throughout Atlantic Canada in summer; year-round on the coast.

Green-winged Teal

Anas crecca

Length: 30–41 cm
Wingspan: 51–58 cm

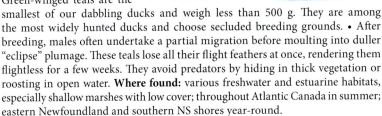

Green-winged teals are the smallest of our dabbling ducks and weigh less than 500 g. They are among the most widely hunted ducks and choose secluded breeding grounds. • After breeding, males often undertake a partial migration before moulting into duller "eclipse" plumage. These teals lose all their flight feathers at once, rendering them flightless for a few weeks. They avoid predators by hiding in thick vegetation or roosting in open water. **Where found:** various freshwater and estuarine habitats, especially shallow marshes with low cover; throughout Atlantic Canada in summer; eastern Newfoundland and southern NS shores year-round.

Common Eider

Somateria mollissima

Length: 50–68 cm
Wingspan: 88–106 cm

Insulative down keeps the common eider warm in cold Atlantic waters and is used to line the bird's nest. The down is also prized by people, and down gathering still takes place in Newfoundland and Québec. • This deep-sea diver preys on molluscs and crustaceans on the ocean floor at depths reaching 45 m. • Small summer flocks grow into rafts of several hundred eiders in winter. Raft diving is amusing to observe—the birds on the edge lead a group dive with the adjacent birds following in sequence, much like fans at a sporting event doing "the wave" in the stands of an arena. **Where found:** shallow coastal waters year-round; occasionally on large, freshwater lakes; Gulf of St. Lawrence, eastern NS, Bay of Fundy and northern Newfoundland.

Common Merganser

Mergus merganser

Length: 56–70 cm
Wingspan: 86 cm

To take off, the common merganser must run along the surface of the water, beating its heavy wings to gain sufficient lift, but once up and away, it flies arrow-straight and low over the water. • This large duck will nest in a tree cavity, occasionally on the ground, on a cliff ledge or in a large nest box, usually close to water. In winter, any source of open water with a fish-filled shoal will support good numbers of these skilled divers. **Where found:** large rivers and deep lakes; throughout most of Atlantic Canada in summer; southern Maritimes year-round.

Ring-necked Pheasant

Phasianus colchicus

Length: *Male:* 75–90 cm; *Female:* 50–55 cm
Wingspan: *Male:* 78 cm; *Female:* 70 cm

Introduced in the late 1800s, this Asian bird has endured many pressures. Populations have had to be continually replenished by hatchery-raised young, not only because this species is hunted, but also because of diminishing habitat, intensive farming practices and our harsh winters. Unlike native grouse, this pheasant lacks feathered legs and feet for insulation and cannot survive on native plants, depending instead on grain and corn crops. **Where found:** shrubby grasslands, urban parks, woodlots, hayfields and croplands; year-round in the Maritimes.

Ruffed Grouse

Bonasa umbellus

Length: 38–48 cm
Wingspan: 55 cm

Every spring, and occasionally in autumn, the male ruffed grouse "drums" to proclaim his territory. He struts along a fallen log with his tail fanned and his neck feathers ruffed, beating the air with accelerating wing strokes. • In winter, scales grow out along the sides of the ruffed grouse's feet, creating temporary "snowshoes." **Where found:** hardwood and mixed forests, riparian woodlands; young, second-growth stands with birch and aspen; throughout Atlantic Canada (introduced in Newfoundland). **Also known as:** partridge.

Willow Ptarmigan

Lagopus lagopus

Length: 38 cm
Wingspan: 61 cm

This Arctic bird was introduced to the warmer climes of Nova Scotia in the late 1960s. • In summer, this bird has white wings and mottled plumage, which transitions to all white with a black tail tip in winter. • The male willow ptarmigan performs raucous courtship calls and boisterous territorial displays. **Where found:** high, rocky slopes, tundra with willow and alder shrubs and other vegetation; NL, introduced into NS.

Common Loon

Gavia immer

Length: 75–88 cm
Wingspan: 1.2–1.5 m

When the haunting call of the common loon
pierces a still evening, cottagers know that summer has begun. This loon actually
has several different calls; individuals have a laughing distress call, separated
pairs seem to wail _Where aaare you?_ and groups give soft, cohesive hoots as they
fly. • Most birds have hollow bones, but the loon has solid bones that reduce its
buoyancy and allow it to dive to depths of 55 m. **Where found:** _Breeding:_ large
lakes and rivers, often with vegetated islands or even muskrat lodges to nest on;
also coastal waters; throughout Atlantic Canada. _Winter:_ ocean bays and headlands; far eastern coastlines.

Pied-billed Grebe

Podilymbus podiceps

Length: 30–38 cm
Wingspan: 57 cm

Very rare in winter, when the larger,
more interesting-looking horned grebe
(_Podiceps auritus_) and red-necked grebe (_Podiceps grisegena_) can be seen on the
Atlantic coast, the pied-billed grebe is nonetheless our most common and only
nesting grebe. It builds a floating nest in the middle of a pond so it can watch
for predators. If a threat is present, the grebe will cover its eggs with vegetation
and then slip into the water with only its eyes and nostrils above the surface.
Where found: ponds, marshes and backwaters with sparse emergent vegetation;
Gaspé Peninsula, NB, NS, PEI and the southern tip of Newfoundland. **Also
known as:** hell-diver.

Northern Gannet

Morus bassanus

Length: 89–97 cm
Wingspan: 1.8 m

If you are boating offshore, keep your eyes open for these
amazing, skydiving birds. From over 30 m in the air,
northern gannets tuck in their wings and plunge
torpedo-style into the water in pursuit of fish.
They have reinforced skulls to cushion the impact.
• A breeding pair of gannets mates for life and performs elaborate pair-bonding
rituals during nesting that involve much preening, bowing and sky pointing as
well as raising their wings and spreading their tails. **Where found:** coastal and open
waters most of the year; nests on coastal cliffs in large colonies; throughout offshore
Atlantic Canada. **Also known as:** solan goose (in Europe).

Double-crested Cormorant

Phalacrocorax auritus

Length: 66–81 cm
Wingspan: 1.3 m

The double-crested cormorant looks like a bird but smells and swims like a fish. With a long, rudder-like tail, excellent underwater vision, sealed nostrils for diving and "wettable" feathers that lack oil glands, this bird has mastered the underwater world. After a dive, a cormorant often perches with its wings partially spread, drying out its feathers. • The male's "ear tufts" are only visible in breeding season. **Where found:** large lakes and large, meandering rivers; nests in colonies on platforms of sticks and guano on islands or in trees; along most coastlines in summer or migration; rare in winter. **Also known as:** shag (in Europe).

Great Cormorant

Phalocrocorax carbo

Length: 91 cm
Wingspan: 1.6 m

The great cormorant is very similar to the double-crested cormorant, but it is larger, with a larger, bulkier bill, a heavier jaw and more white in its plumage, including a white patch under its yellow throat pouch. In breeding plumage, wispy, white feathers appear on the head and the yellow throat pouch turns dark whereas that of the double-crested cormorant turns orange in breeding. • Great cormorants nest in colonies on cliffs or in trees, and Atlantic Canada is an important breeding range for this species in North America. **Where found:** coastal estuaries and some freshwater lakes and rivers; throughout Atlantic Canada in breeding season. **Also known as:** great black cormorant.

American Bittern

Botaurus lentiginosus

Length: 70 cm
Wingspan: 1.1 m

When an American bittern hears or sees you approach, it stands completely still with its bill pointing skyward, its vertically streaked, brown plumage blending in with the reeds and rushes. You are more likely to hear a bittern than spot one; in spring, the marshlands resonate with its deep, booming mating call. **Where found:** marshes, wetlands and lake edges with reeds, rushes and sedges; throughout all but northern QC and Labrador in summer. **Also known as:** marsh-pumper.

Great Blue Heron

Ardea herodias

Length: 1.3–1.4 m
Wingspan: 1.8 m

The long-legged great blue heron has a stealthy, often motionless hunting strategy. It waits for a fish or frog to approach, spears the prey with its bill, then flips its catch into the air and swallows it whole. This heron usually hunts near water, but it also stalks fields and meadows in search of rodents. • Great blue herons settle in communal treetop nests called rookeries, and nest width can reach over 1 m. **Where found:** forages along the edges of rivers, lakes and marshes; also in fields and wet meadows; Gulf of St. Lawrence, NS, PEI and NB in summer.

Green Heron

Butorides virescens

Length: 46 cm
Wingspan: 66 cm

The intelligent green heron uses bait to catch fish—it drops small debris, such as bits of vegetation or a feather, onto the water's surface to attract its prey. Standing still among the weedy wetland vegetation, the heron goes unnoticed by the fish until the bird strikes. This heron is also a skilled frog hunter. • The metallic green feathers are inconspicuous unless the light hits them at the right angle, making them shimmer. **Where found:** freshwater marshes, lakes and streams with plenty of vegetation; local breeder in southern NB, otherwise possibly north and east of this range in migration.

Black-crowned Night-Heron

Nycticorax nycticorax

Length: 58–66 cm
Wingspan: 1.1 m

When dusk's long shadows shroud the marshes, black-crowned night-herons arrive to hunt in the marshy waters. These herons crouch motionless, using their large, light-sensitive eyes to spot prey lurking in the shallows. Look for them in summer, between dawn and dusk, as they fly between nesting and feeding areas. **Where found:** shallow cattail and bulrush marshes, lakeshores and along slow-flowing rivers; also on sheltered coastlines along the Gaspé Peninsula, NB and southern NS; throughout much of the southern Maritimes in migration.

Osprey

Pandion haliaetus

Length: 56–64 cm
Wingspan: 1.7–1.8 m

The osprey is the provincial bird of Nova Scotia. • While hunting for fish, this large, powerful raptor hovers in the air before hurling itself in a dramatic headfirst dive. An instant before striking the water, it rights itself and thrusts its feet forward to grasp its quarry. • Ospreys build bulky nests on high, artificial structures such as communication towers and utility poles or on buoys and channel markers over water, where the pair tends to 2 to 3 chicks. **Where found:** lakes and slow-flowing rivers and streams; estuaries and bays in migration; throughout Atlantic Canada in summer.

Bald Eagle

Haliaeetus leucocephalus

Length: 76–109 cm
Wingspan: 1.7–2.4 m

While soaring high in the air, a bald eagle can spot fish swimming underwater and small rodents scurrying through the grass. • This eagle does not mature until its fourth or fifth year—only then does it develop the characteristic white head and tail plumage. • Bald eagles mate for life and renew their pair bonds by adding sticks to the same nest each year. Nests can be up to 5 m in diameter and are the largest of any North American bird. **Where found:** coastal areas, estuaries, large lakes, river valleys and farmlands; year-round in southern coastal Newfoundland and much of the Maritimes; also in QC and NL in summer.

Northern Harrier

Circus cyaneus

Length: 41–61 cm
Wingspan: 1.1–1.2 m

The courtship flight of the northern harrier is a spectacle worth watching in spring. The male climbs almost vertically in the air, then stalls and plummets in a reckless dive toward the ground. At the last second, he saves himself with a hairpin turn that sends him skyward again. • Britain's Royal Air Force named the Harrier aircraft after this raptor because of the bird's impressive manoeuvrability. **Where found:** open country including fields, wet meadows, cattail marshes, bogs and croplands; nests on the ground, usually in tall vegetation; along much of our coastline in summer; year-round only in southern NS.

Sharp-shinned Hawk

Accipiter striatus

Length: *Male:* 25–30 cm; *Female:* 30–36 cm
Wingspan: *Male:* 51–61 cm; *Female:* 61–71cm

After a successful hunt, the small sharp-shinned hawk often perches on a favourite "plucking post," holding its meal in its razor-sharp talons. This woodland raptor preys almost exclusively on small birds. • Short, rounded wings, a long, rudder-like tail and flap-and-glide flight allow this hawk to manoeuvre through the forest at high speed. • As it ages, the sharp-shinned hawk's bright yellow eyes become red. **Where found:** dense to semi-open coniferous forests and large woodlots; occasionally along rivers and in urban areas; may visit backyard bird feeders to prey on sparrows and finches in winter; in summer, throughout all but northern NL, except possibly during migration.

Northern Goshawk

Accipiter gentilis

Length: *Male:* 53–58 cm; *Female:* 58–64 cm
Wingspan: 1–1.2 m

This forest raptor navigates through the trees in swift and agile pursuit of its prey, which includes birds and small mammals. • The northern goshawk will aggressively protect its nest from any perceived threats from predators or even innocent passersby with aerial, dive-bombing assaults accompanied by a deafening attack screech. **Where found:** mature coniferous and deciduous forests and mixed woodlands; forest edges, parkland and farmland in winter; throughout Atlantic Canada year-round.

Broad-winged Hawk

Buteo platypterus

Length: 41 cm
Wingspan: 86 cm

This hawk is typically seen perched on fence posts and power lines watching for prey, which includes small birds and mammals, reptiles, amphibians and large insects. • Slow, regular wingbeats, white underwings with pointed, black wing tips and the single, broad, white band on the dark tail help identify this raptor in flight. • The broad-winged hawk is found in our region only in summer; it winters in Florida, Texas or South America. **Where found:** in woodlands and along forest edges; southern Gulf of St. Lawrence and the Maritimes in summer.

Red-tailed Hawk

Buteo jamaicensis

Length: *Male:* 46–58 cm; *Female:* 51–64 cm
Wingspan: 1.2–1.5 m

Spend a summer afternoon in the country and you will likely see a red-tailed hawk perched on a fence post or soaring on a thermal. • Courting red-tails will sometimes dive at one another, lock talons and tumble toward the earth, breaking away at the last second to avoid crashing into the ground. • The red-tailed hawk's piercing call is often paired with the image of an eagle in TV commercials and movies. **Where found:** open country with some trees; also roadsides and woodlots; often flies above cities; throughout the Maritimes and the southern Gulf of St. Lawrence.

American Kestrel

Falco sparverius

Length: 19–20 cm
Wingspan: 51–61 cm

The colourful American kestrel, formerly known as "sparrow hawk," is not shy of human activity and is adaptable to habitat change. This small falcon has benefited from the grassy roadway margins that provide habitat for grasshoppers, which make up most of its diet, and other small prey such as mice. **Where found:** along rural roadways, perched on poles and telephone wires; agricultural fields, grasslands, riparian woodlands, woodlots, forest edges, bogs, roadside ditches and grassy highway medians; in summer, throughout the Maritimes and parts of QC, including Anticosti I., and Newfoundland.

Merlin

Falco columbarius

Length: 25–30 cm
Wingspan: 58–66 cm

The merlin is a powerful flyer and deftly hunts on the wing, catching bird prey such as sparrows and pigeons. • This species has been of concern because of declines in its population, which were thought to be related to pesticide use. • The merlin breeds and nests in our region, laying 5 to 6 eggs in a tree cavity or on a rock ledge without a nest or any sort of lining. **Where found:** shrublands, coniferous forests adjacent to open fields and sometimes in suburban areas; throughout Atlantic Canada in summer; year-round Maritime resident when prey is abundant. **Also known as:** pigeon hawk.

Virginia Rail

Rallus limicola

Length: 24 cm
Wingspan: 33 cm

Although it is secretive like all rails, the Virginia rail likes to sing. In breeding season, which is when the Virginia rail is found in our region, its repertoire includes a territorial song—a two-phrase *tik-tik, tik-tik-tik* or *kid kid kidick kidick*—as well as a *tik tik tik turrr*, which may have another purpose. Year-round, this bird's call is a descending, accelerating series of notes that has been described as a raspy *oink*. **Where found:** among sedges and reeds in wet grasslands, wetlands and coastal and inland marshes; throughout the southern Gulf of St. Lawrence and the Maritimes in summer.

Sora

Porzana carolina

Length: 20–25 cm
Wingspan: 36 cm

The sora has a small body and large, chicken-like feet. Even without webbed feet, this unique creature swims quite well over short distances. • Two rising *or-Ah or-Ah* whistles followed by a strange, descending whinny indicate that a sora is nearby. This secretive bird is hard to spot because it prefers to remain hidden in dense marshland, but it will occasionally venture into the shallows. **Where found:** wetlands with abundant emergent vegetation; also grain fields; throughout Atlantic Canada in summer; in saltwater marshes during migration.

Semipalmated Plover

Charadrius semipalmatus

Length: 18 cm
Wingspan: 38 cm

The semipalmated plover is one of those cute little birds that run back and forth chasing and being chased by the waves. They do not do it for fun, though; they are running to snatch tiny prey such as aquatic worms, crustaceans and insects that are thrown ashore by the surf. • This little bird ranges between breeding grounds in the Arctic and wintering grounds as far south as Patagonia. **Where found:** mudflats, lakeshores, estuaries, beaches and sandbars; breeds on the tundra along northern coastline, primarily the northeastern coasts of QC, NL and NS, but scattered breeding populations are found throughout the Maritimes.

Killdeer

Charadrius vociferus

Length: 23–28 cm
Wingspan: 60 cm

The killdeer is a gifted actor, well known for its "broken wing" distraction display. When an intruder wanders too close to its ground nest, the killdeer utters piteous cries while dragging a wing and stumbling about as if injured. Most predators take the bait and follow, and once the killdeer has lured the predator far away from its nest, it miraculously recovers from the injury and flies off with a loud call. **Where found:** open, wet meadows, lakeshores, sandy beaches, mudflats, gravel streambeds and golf courses; throughout the region in summer.

Spotted Sandpiper

Actitis macularius

Length: 18–20 cm
Wingspan: 38 cm

The female spotted sandpiper diligently defends her territory, mates, lays her eggs and leaves the male to tend the clutch. Only about one percent of birds display this unusual breeding strategy, which is known as "polyandry." She may mate with several different males, lay up to 4 clutches and produce 20 eggs in one summer. **Where found:** shorelines, gravel beaches, drainage ditches, swamps and sewage lagoons; occasionally cultivated fields; throughout Atlantic Canada in summer.

Greater Yellowlegs

Tringa melanoleuca

Length: 29–38 cm
Wingspan: 58 cm

The greater yellowlegs and lesser yellowlegs (*T. flavipes*) are medium-sized sandpipers with very similar plumages and very yellow legs and feet. The species differ subtly, and a solitary yellowlegs is difficult to identify until it flushes and utters its distinctive cry—the greater yellowlegs peeps 3 times, and the lesser yellowlegs peeps twice. As its name suggests, the greater yellowlegs is the larger species, and it has a slightly upturned, longer bill that is about one and a half times the width of its head. **Where found:** any type of shallow wetland, whether freshwater, brackish or salt; flooded agricultural fields; migrates through the Maritimes and southern QC to summer breeding ranges in NL and northern QC.

Ruddy Turnstone

Arenaria interpres

Length: 24 cm
Wingspan: 53 cm

This shorebird breeds on the Arctic tundra and
migrates through our area in autumn, headed for the warmer
coastlines between New England and the Gulf of Mexico.
A rather dignified shorebird, it prefers to casually walk among the rocks and
mudflats rather than frantically running to and fro with the waves. • Turnstones
are likely named for their habit of picking through rocks in search of crustaceans,
worms and other invertebrates. **Where found:** coastal jetties and mudflats; along
the Atlantic coastline during spring and autumn migrations.

Sanderling

Calidris alba

Length: 20 cm
Wingspan: 43 cm

The sanderling chases the waves in and out, snatching up aquatic invertebrates before they are swept back
into the water. On shores where wave action is limited,
it resorts to probing mudflats for a meal of molluscs and
insects. • To keep warm, a sanderling will seek the company of other roosting
shorebirds. It will also stand with one leg tucked up, a posture that conserves body
heat. **Where found:** sandy and muddy shorelines, cobble and pebble beaches, spits,
lakeshores, marshes and reservoirs; coastal winter resident in southern NS;
all along the Atlantic coastline during spring and autumn migrations.

White-rumped Sandpiper

Calidris fuscicollis

Length: 18 cm
Wingspan: 36 cm

This tundra-nesting shorebird frequents our coastlines
in large, conspicuous flocks in late spring and
autumn. Its long migrations take it as far as South
America—one of the longest migrations of any North
American bird in distance as well as duration, which is about a month. Food-rich
shorelines such as ours are important because the birds need to maintain a healthy
weight to endure the long migration, particularly en route north, when they need to
arrive in good condition to nest. • Although difficult to distinguish from other
small shorebirds, the white-rumped sandpiper has a unique *tzeep*! call. **Where
found:** mudflats, beaches and lakeshores; throughout Atlantic Canada.

89

Dunlin

Calidris alpina

Length: 22 cm
Wingspan: 43 cm

The dunlin is our most common winter shorebird and can often be seen in very large flocks at tide-line roosts. Its winter plumage, when we observe it, is unfortunately rather drab, but by spring, just before the bird migrates, you may see its more attractive summer plumage—a lovely russet on its back and a bold, black belly. **Where found:** tidal and saltwater marshes, estuaries, lagoon shorelines, open, sandy ocean beaches, flooded fields and muddy wetlands; all along the Atlantic coastline during migration; coastal winter resident and migrant through the interior; coastal winter resident in southern NS.

Wilson's Snipe

Gallinago delicata

Length: 27 cm
Wingspan: 46 cm

When flushed from cover, snipes perform a series of aerial zigzags to confuse predators. Because of this habit, hunters who were skilled enough to shoot these birds became known as "snipers," a term later adopted by the military.
• Courting snipes make an eerie, winnowing sound, like a rapidly hooting owl. The sound is produced by the male's specialized outer tail feathers, which vibrate rapidly in the air as he performs daring, headfirst dives high above a wetland. **Where found:** cattail and bulrush marshes, sedge meadows, poorly drained floodplains, bogs and fens; also willow and dogwood tangles; throughout Atlantic Canada in summer.

Black-headed Gull

Chroicocephalus ridibundus

Length: 38–41 cm
Wingspan: 90–102 cm

This gull is typically seen in winter when it does not have the black head of its namesake, but it does have a prominent black spot behind its eye that helps in identification, and the legs and bill are yellowish. The black-headed gull does, however, have a breeding population in Newfoundland and is commonly observed in other Maritime provinces in its breeding plumage with the black head, red bill and legs. **Where found:** coastal in winter; nests in freshwater marshes and coastal salt marshes; winter resident throughout the eastern Maritimes with breeding populations in Newfoundland. **Also known as:** *Larus ridibundus*.

Herring Gull

Larus argentatus

Length: 58–66 cm
Wingspan: 1.2 m

The voracious appetite of this gull as it scavenges
the beaches for dead fish, crustaceans and any other
carrion benefits us by helping keep the beaches clean.
However, this gull is not widely appreciated for stealing meals from fishermen,
eating fish placed on fields for fertilizer and, in particular, for its habit of dropping
shellfish from high in the air to crack open the shells, often denting car roofs.
• The herring gull is one of the most abundant gulls in our region. **Where found:**
coastlines, rivers and lakes; throughout Atlantic Canada year-round.

Great Black-backed Gull

Larus marinus

Length: 76 cm
Wingspan: 1.6 m

A non-discriminating scavenger, aggressive
predator and ruthless food thief, stealing
meals from other birds, the great black-
backed gull has a diverse diet of fish, small mammals,
the chicks of other seabirds and garbage. • This gull
builds its nest on a mat of vegetation on the ground, but
it may also be found in colonies on rocky ledges. One of
the largest nesting colonies is located at Lake George near Yarmouth, Nova Scotia.
It is gregarious in winter, when it flocks with other gulls species. **Where found:**
coastlines, large lakes, urban areas and garbage dumps; throughout Atlantic
Canada year-round.

Common Tern

Sterna hirundo

Length: 31–37 cm
Wingspan: 76 cm

This colony nester is a gregarious presence on our coastlines and wetlands in summer. It lays its eggs in a scrape or on
a platform of vegetation on the ground, a precarious location if
on a heavily trafficked beach—always watch your step in spring. • This tern's
breeding plumage shows a dark cap and red bill, whereas nonbreeding juveniles
have a pale bill and a thinning of the black cap at the front of the head, giving the
appearance of a receding hairline. • Terns dive like little torpedoes into the water
to catch fish and other small aquatic animals. **Where found:** lakes, islands,
marshes and coastlines; throughout most of Atlantic Canada in summer.

Razorbill

Alca torda

Length: 43 cm
Wingspan: 66 cm

In the water, the razorbill can dive to depths of 60 m to catch fish or grab crustaceans and invertebrates from the seabed. It also supplements its captures with fish pirated from other birds. • Like other alcids, the razorbill moults during nesting and mating, becoming temporarily flightless. • The female razorbill lays a single egg on a rocky ledge. • The razorbill is likely the closest living relative to the extinct great auk (*Pinguinus impennis*). **Where found:** open ocean or rocky coastal cliffs; throughout offshore waters; in large numbers on the Grand Banks off Newfoundland; breeds mainly in south-eastern Labrador; winters offshore from southern NS. **Also known as:** tinker, noddy.

Black Guillemot

Cepphus grylle

Length: 30 cm
Wingspan: 53 cm

These alcids are easy to watch from shore while they feed on crustaceans and other invertebrates that are churned up in the coastal currents. Guillemots are skilled divers and underwater swimmers, and also prey on small fish and eels. • The black breeding plumage for which the species is named becomes white in winter. **Where found:** inshore and along rocky coasts; northern Gulf of St. Lawrence, lower Bay of Fundy and the Atlantic coast. **Also known as:** sea pigeon.

Atlantic Puffin

Fratercula arctica

Length: 32 cm
Wingspan: 53 cm

The Atlantic puffin is the provincial bird of Newfoundland and Labrador. • Young puffins spend the first years of their lives at sea on the open water without touching land. • This bird's gorgeous, colourful bill takes up to 5 years to develop in size and form. • The puffin dives and swims to catch fish, and can hold over a dozen small fish crosswise in its bill. **Where found:** from the Maine–NB border north to southeastern Labrador, mainly in southeastern Newfoundland; established breeding colonies are found in turf and among boulders on the tops and sides of coastal cliffs; in winter, some migrate along the coast as far south as Massachusetts. **Also known as:** sea parrot.

Rock Pigeon

Columba livia

Length: 30–33 cm
Wingspan: 71 cm

The rock pigeon is likely a descendant of a Eurasian
bird that was first domesticated in about 4500 BC.
Both Caesar and Napoleon used rock pigeons
as message couriers. European settlers introduced the species
to North America in the 17th century, and today this bird is
familiar to almost everyone. • No other "wild" bird varies as much in colouration,
a result of semi-domestication and extensive inbreeding over time. **Where found:**
urban areas, railway yards and agricultural areas; high cliffs often provide habitat
in the wild; throughout the region. **Also known as:** rock dove.

Mourning Dove

Zenaida macroura

Length: 27–33 cm
Wingspan: 45 cm

This dove is one of the most abundant native birds in North
America. Its numbers and range have increased as human
development has created more open habitats and food
sources, such as waste grain and bird feeders. • The
mourning dove's soft cooing, which filters through
broken woodlands and suburban parks, is often con-
fused with the sound of a hooting owl. • The female lays only 2 eggs a time but
may produce up to 6 broods each year—more than any other native bird. **Where
found:** open and riparian woodlands, forest edges, agricultural and suburban
areas and open parks; throughout the Maritimes and southern QC in summer.

Black-billed Cuckoo

Coccyzus erythropthalmus

Length: 28–31 cm
Wingspan: 38 cm

The insectivorous black-billed cuckoo
enjoys munching on caterpillars, particularly tent caterpillars. • The cuckoo sings its name,
cu-cu-cu cu-cu-cu, which superstition says predicts rain. • The yellow-
billed cuckoo (*C. americanus*) is the American cousin to our cuckoo, though it can
be a rare southern Maritime vagrant at the northern edge of its range, which is
almost precisely the Canada–U.S. border. **Where found:** forest edges, thickets,
woodlands, scrublands and along streams; throughout the Maritimes and the southern St. Lawrence and Gaspé regions. **Also known as:** rain bird.

Great Horned Owl

Bubo virginianus

Length: 45–64 cm
Wingspan: 0.9–1.5 m

This highly adaptable and superbly camouflaged hunter has sharp hearing and powerful vision that allow it to hunt by night and day. It will swoop down from a perch onto almost any small creature that moves. The leading edge of the flight feathers is fringed rather than smooth, which interrupts airflow over the wing and allows the owl to fly noiselessly. • The great horned owl has a poor sense of smell, which might explain why it is the only consistent predator of skunks. **Where found:** fragmented forests, fields, riparian woodlands, suburban parks and small, mature woodlots with agricultural land nearby; throughout the region.

Snowy Owl

Bubo scandiacus

Length: 58 cm
Wingspan: 1.3 m

There are few more beautiful winter sights than that of a snowy owl swooping over a frosty landscape on a moonlit night. We are graced with this owl's presence for a few months each year as it enjoys an escape from its frozen Arctic habitat. • Feathered to the toes, the snowy owl can endure temperatures that send other owls to the woods to seek shelter. It is active day and night, hunting small rodents, especially lemmings, as well as sea and water birds. **Where found:** perches on the ground, fence posts, low stumps and buildings; throughout the region; NS appears to be the southern extent of its East Coast range.

Barred Owl

Strix varia

Length: 51 cm
Wingspan: 1.1 m

Our most common owl, the barred owl hoots a distinctive, rhythmic call, described as *Who cooks for you, who cooks for you all?* It is recognizable for the distinct, dark rings on its facial disc, but more interesting to note are the dark, nearly completely black eyes. Only the barn owl (*Tyto alba*) has these same dark eyes; all the rest of our owls have yellow eyes. • The barred owl nests in tree cavities and is dependent upon old-growth forests for habitat. **Where found:** old-growth, mixed woodlands, dense coniferous forests and swampy areas; throughout Atlantic Canada.

Long-eared Owl

Asio otus

Length: 38 cm
Wingspan: 92 cm

A friend of the farmer but a terror to rats and mice, the long-eared owl is credited both for its diet, of which 80 to 90 percent is made up of injurious rodents, and for its distaste for domestic poultry. It hunts by night, and by day it roosts deep in the seclusion of the forest. • The long "ears" are really just tufts of feathers. • To scare off an intruder, this owl expands its air sacs, puffs its feathers and spreads its wings. **Where found:** forest and woodland edges, open fields and riparian stands; southern St. Lawrence region, the Gaspé Peninsula and the Maritimes.

Short-eared Owl

Asio flammeus

Length: 38 cm
Wingspan: 96 cm

The short-eared owl nests on the ground, sometimes concealed by thick grasses or shrubs, but often quite exposed. Normally silent, it will cry out with crow-like squawks and screeches to deter any potential threat that approaches too closely to its nest. • The short-eared owl is a skilled crepuscular mouser, flying over the countryside at dusk in search of rodent prey. **Where found:** open countryside, marshes, tundra and fields; throughout Atlantic Canada in summer, ranging from Baffin I. south through New England.

Northern Saw-whet Owl

Aegolius acadicus

Length: 20 cm
Wingspan: 42 cm

A nocturnal owl, the saw-whet is difficult to observe and is more frequently heard—it has various calls that sound like whistles, screeches and barks. The name "saw-whet" refers to one of this owl's calls, which is similar to the metallic sound of a saw being sharpened. • The saw-whet nests in hollow trees, often reusing the nest of a northern flicker, or sometimes in a nest box. By day, it roosts in a tree in dense forest. **Where found:** coniferous or mixedwood forests, swamps and tamarack bogs; throughout the Maritimes and the Gaspé Peninsula year-round; northern St. Lawrence region in summer.

Common Nighthawk

Chordeiles minor

Length: 20–25 cm
Wingspan: 58–66 cm

The common nighthawk, like all nightjars, has adapted to catch insects in midair—its large, gaping mouth is surrounded by feather shafts that funnel insects into its bill. This bird can eat more than 2600 insects in one day, including mosquitoes, black flies and flying ants. • In an energetic courting display, the male nighthawk dives, then swerves skyward, making a hollow booming sound with its wings. **Where found:** *Breeding:* forest openings, bogs, rocky outcroppings and gravel rooftops; southern Labrador and rarely Newfoundland. *In migration:* near water or any area with large numbers of flying insects; throughout the region except PEI.

Ruby-throated Hummingbird

Archilochus colubris

Length: 8 cm
Wingspan: 11 cm

Several dozen tiny wingbeats per second allow this aerial extremist to manoeuvre forward and backward and even hover like a helicopter. At full tilt, this nickel-weight speedster has a heart rate of over 1000 beats per minute. • Sugarwater or nectar feeders can attract dozens of ruby-throated hummingbirds, which will joust with other hummers for access. • The female lays 2 jellybean-sized eggs within a nest the size of half a walnut shell, woven with spiderwebs and shingled in lichen. **Where found:** open, mixed woodlands, wetlands, gardens and backyards; breeds throughout the Maritimes, the St. Lawrence region and the Gaspé Peninsula.

Belted Kingfisher

Megaceryle alcyon

Length: 28–36 cm
Wingspan: 50–53 cm

From a bare-branch perch over a productive pool, a belted kingfisher will plunge headfirst into the water, snatch up a fish or frog, flip it into the air, and then swallow it headfirst. Nestlings are able to swallow small fish whole when they are only 5 days old. • With a reddish band across her belly, the female kingfisher is more colourful than her mate. **Where found:** rivers, large streams, lakes, marshes and beaver ponds, especially near exposed soil banks, gravel pits or bluffs; throughout the region in summer. **Also known as:** *Ceryle alcyon*.

Downy Woodpecker

Picoides pubescens

Length: 15–18 cm
Wingspan: 30 cm

A bird feeder well stocked with peanut butter and sunflower seeds may attract a pair of downy woodpeckers to your backyard. These approachable little birds are more tolerant of human activity than are most other species, and they visit feeders more often than the larger, more aggressive hairy woodpeckers (*P. villosus*). • The downy woodpecker's white outer tail feathers have several dark spots, whereas the hairy's are pure white. **Where found:** any wooded environment, especially deciduous and mixed forests and areas with tall, deciduous shrubs; throughout the region year-round except Labrador.

Black-backed Woodpecker

Picoides arcticus

Length: 24 cm
Wingspan: 43 cm

Rather than knocking holes in trees, and perhaps to avoid rattling its skull too much, the black-backed woodpecker tends to flake bark off old trees in search of larvae and insects hiding beneath. It requires old-growth forest habitat to find trees full of invertebrates as well as hollows in which to nest. It is also attracted to burned areas. **Where found:** coniferous, old-growth forests: throughout Atlantic Canada year-round.

Northern Flicker

Colaptes auratus

Length: 32 cm
Wingspan: 50 cm

The northern flicker scours the ground and tree trunks in search of invertebrates, particularly ants, which it squashes and then uses to preen itself. The formic acid in the ants' bodies kills small parasites on the bird's skin and feathers. • There are 2 races of northern flicker: the yellow-shafted flicker of eastern North America has yellow underwings and undertail coverts, whereas those of the red-shafted flicker of the West are reddish. **Where found:** open woodlands, forest edges, fields, meadows, beaver ponds and other wetlands in summer; throughout Atlantic Canada.

Pileated Woodpecker

Dryocopus pileatus

Length: 42–50 cm
Wingspan: 69 cm

This crow-sized bird, the sixth-largest woodpecker in the world, is an unforgettable sight if you are fortunate enough to see one. Despite its size, noisy wood-pecking habits and maniacal breeding calls, it is quite elusive. Large, distinctively oval-shaped nest holes reveal its presence, as do trees that look as if someone has taken an axe to them. • Wood ducks, American kestrels, owls and even flying squirrels nest in abandoned pileated woodpecker nest holes. **Where found:** large, mature forests; year-round throughout the Maritimes, Gaspé Peninsula and southern St. Lawrence region.

Olive-sided Flycatcher

Contopus cooperi

Length: 18–20 cm
Wingspan: 33 cm

The olive-sided flycatcher's upright, attentive posture contrasts with its comical song: *quick-three-beers! quick-three-beers!* Like a dutiful parent, this flycatcher changes its tune during nesting, when it more often produces an equally enthusiastic *pip-pip-pip.* • The olive-sided flycatcher builds a nest of twigs bound with spider silk, high in a conifer, usually on a branch far from the trunk. • This species' numbers are declining, and it is listed as a federal species of concern. **Where found:** semi-open, mixed and coniferous forests near water, burned areas and wetlands; throughout Atlantic Canada in summer, except northern Labrador.

Alder Flycatcher

Empidonax alnorum

Length: 13–15 cm
Wingspan: 20 cm

The alder flycatcher forages for insects by catching them in midair. It is constantly on the move, never seeming to rest as it flits from shrub to shrub, catching a morsel on each pass. • This bird's nest is made of grasses and affixed to a low shrub. It adorns its nest with plant down or long grass stems, which do not seem to serve any purpose other than exterior decoration. **Where found:** brushy habitats near bogs, swamps and wetlands; also birch and alder thickets; throughout Atlantic Canada in summer, except northern Labrador.

Eastern Kingbird

Tyrannus tyrannus

Length: 22 cm
Wingspan: 36 cm

As its scientific name suggests, the eastern kingbird is somewhat of a tyrant among birds, harassing and mobbing crows, hawks and even humans that enter its territory. It is no friend of insects, either, for which this bird has a voracious appetite.
• You will often see kingbirds perched on fencelines or utility wires along roadsides. **Where found:** fields and agricultural landscapes, woodland clearings and near water; in summer, throughout the Maritimes, southern QC, the Gaspé Peninsula and the southern St. Lawrence region.

Northern Shrike

Lanius excubitor

Length: 25 cm
Wingspan: 36–38 cm

The northern shrike visits our region each winter in unpredictable and highly variable numbers. • One of the most vicious predators in the bird world, this bird relies on its sharp, hooked bill to catch and kill small birds and rodents, which it spots from treetop perches. Its tendency to impale its prey on thorns and barbs for later consumption has earned it the name "butcher bird." • Shrikes are the world's only truly carnivorous songbirds. **Where found:** semi-open country, scrublands and low-elevation orchards, farmlands and ranches; migrants often appear among dunes or in coastal scrub.

Red-eyed Vireo

Vireo olivaceus

Length: 15 cm
Wingspan: 25 cm

One of the most common breeding birds in our area, the red-eyed vireo is difficult to spot high in the canopy where it prefers to forage. • Virtuoso male vireos have an impressive repertoire of songs, with a record 21,000 tallied in a single day and dozens of phrases expressed each minute. Most of the songs are likely improvised, one-time performances for lucky listeners. **Where found:** deciduous or mixed woodlands; throughout Atlantic Canada except small ranges in southern Newfoundland and parts of Labrador. **Also known as:** preacher bird.

Gray Jay

Perisoreus canadensis

Length: 29 cm
Wingspan: 45 cm

Less boisterous than other jays, the gray jay is quite amicable, cheerfully approaching campsites and even landing on the heads and hands of hikers, politely asking for treats. • In preparation for winter, it covers food that it wants to cache with saliva and sticks it under tree branches. This bird can create thousands of caches per day. • The gray jay's diet includes nuts and berries (sometimes from trail mix), insects, mushrooms, other birds' eggs and nestlings, and carrion. **Where found:** coniferous and boreal forests; year-round throughout Atlantic Canada. **Also known as:** Canada jay, whiskey jack, camp robber.

Blue Jay

Cyanocitta cristata

Length: 28 cm
Wingspan: 41 cm

This loud, striking, well-known bird is the provincial bird of Prince Edward Island. • Blue jays can be quite aggressive when competing for sunflower seeds and peanuts at backyard feeders, rarely hesitating to drive away smaller birds, squirrels or even cats. • Blue jays cache nuts and are very important to the forest ecosystem. In autumn, one jay might bury hundreds of acorns, later forgetting where many were hidden and thus planting scads of oaks. **Where found:** all types of habitats, from dense forests to suburbia; common and widespread year-round throughout southern QC, the Maritimes and Newfoundland.

Common Raven

Corvus corax

Length: 61 cm
Wingspan: 1.4 m

The common raven soars with a wingspan comparable to that of a hawk, travelling along coastlines, over deserts, along mountain ridges and even over the Arctic tundra. Few birds occupy such a large natural range. • From producing complex vocalizations to playfully sliding down snowbanks, this raucous bird exhibits behaviours that many people once thought of as exclusively human. • All corvids are extremely intelligent, some rivalling chimpanzees in problem-solving tests. **Where found:** coniferous and mixed forests and woodlands; also townsites, campgrounds and landfills; throughout Atlantic Canada year-round.

Horned Lark

Eremophila alpestris

Length: 17–20 cm
Wingspan: 30 cm

One way to distinguish a horned lark from a sparrow is by its method of locomotion: horned larks walk, but sparrows hop.
• This bird has a dark tail that contrasts with its light brown body and belly, and it has 2 unique black "horns." This feature will help you to spot the horned lark in its open-country habitat. • In spring, male larks perform impressive, high-speed, plummeting courtship dives. **Where found:** short-grass habitats, farmlands and roadsides in summer; prefers agricultural fields and beaches in winter; throughout Atlantic Canada in summer; southern Maritimes year-round.

Tree Swallow

Tachycineta bicolor

Length: 15 cm
Wingspan: 34 cm

An early bird of spring and one of the last to leave in autumn, tree swallows spend a large part of the year in our region. They are common near old-growth forests, where they find cavities to nest in, but they will also happily take up residence in nest boxes or under barn eaves. • Both parents share the duties of nest building and caring for the nestlings, tirelessly flying to and fro, collecting building materials, catching insects and feeding the hungry babies. **Where found:** old-growth woodlands, near water; throughout Atlantic Canada in summer.

Barn Swallow

Hirundo rustica

Length: 17 cm
Wingspan: 38 cm

Barn swallows are a familiar sight around farmsteads, where they build their mud nests under the eaves of barns and other buildings. It is now almost unheard of for them to nest in natural sites such as cliffs, to which they once were restricted. • The males have elegant, long, forked tails and beautiful colouring. **Where found:** open landscapes, especially in rural and agricultural areas, often near water; throughout the Maritimes, Gaspé Peninsula, St. Lawrence region and southern Newfoundland in summer.

Black-capped Chickadee

Poecile atricapillus

Length: 12–14 cm
Wingspan: 20 cm

The black-capped chickadee is New Brunswick's provincial bird. • A common visitor to backyard feeders, chickadees join the company of kinglets, nuthatches, creepers and small woodpeckers. In spring and autumn, they join mixed flocks of vireos and warblers. • The calling out of its name, *chick-a-dee-dee-dee*, is this bird's most distinctive sound, but it also sings a slow, whistled *swee-tee* or *fee-bee*. **Where found:** deciduous and mixed forests, riparian woodlands and wooded urban parks; also backyard feeders; throughout Atlantic Canada year-round except northern QC and most of Labrador.

Red-breasted Nuthatch

Sitta canadensis

Length: 11 cm
Wingspan: 20 cm

The red-breasted nuthatch has a somewhat dizzying view of the world as it moves down tree trunks headfirst, cleaning up any seeds, insects and nuts that woodpeckers may have overlooked. It is attracted to backyard bird feeders filled with suet or peanut butter. • Nuthatches excavate a cavity nest or use an abandoned woodpecker nest and smear the entrance with sap to keep away ants and other insects that can transmit fungal infections or parasitize nestlings. **Where found:** open woodlands; prefers oak and pine forests; throughout Atlantic Canada year-round except northwestern Labrador.

Winter Wren

Troglodytes hiemalis

Length: 10 cm
Wingspan: 14 cm

The upraised, mottled brown tail of the winter wren blends in well with its habitat of gnarled, upturned roots and decomposing tree trunks. • This tiny bird boldly lays claim to its territory with its call and distinctive, melodious song. It can sustain its song for 10 seconds, using up to 113 tones. • Although the male contributes to raising the family, defending the nest and finding food for the nestlings, he sleeps elsewhere at night, in an unfinished nest. **Where found:** lowland forests and thickets; prefers wet forests; throughout Atlantic Canada in summer but absent from most of Labrador. **Also known as:** Jenny wren; *T. troglodytes*.

Golden-crowned Kinglet

Regulus satrapa

Length: 10 cm
Wingspan: 18 cm

The dainty golden-crowned kinglet is not much bigger than a hummingbird, and when it gleans the forest canopy for insects, berries and sap, it is prone to unique hazards such as perishing on the burrs of burdock plants. • This songbird's perpetual motion and chronic wing flicking can help identify it from a distance. **Where found:** at the tops of the spruces, pines and firs in mature coniferous forests; moves to coastal forests, riparian areas and sometimes urban parks and gardens in migration and winter; year-round throughout the Maritimes, Newfoundland and southeastern QC; north of the St. Lawrence to the tip of Labrador only in summer.

Eastern Bluebird

Sialia sialis

Length: 18 cm
Wingspan: 29 cm

It is a treasured occasion when an eastern bluebird takes up residence in a nest box on your property. Nest boxes placed on fence posts have greatly bolstered bluebird popula-tions. • The male displays gorgeous deep blue and con-trasting warm rufous plumage, and it sings a soft, pleasing, warbling song. The female is duller in colour, and a young bird is heavily spotted below, revealing this species' relationship to the thrushes. **Where found:** agricultural fields and pastures, orchards, fencelines, meadows, open woodlands and forest clearings and edges; southern QC, NB and NS in summer.

Veery

Catharus fuscescens

Length: 18 cm
Wingspan: 28 cm

At twilight, this woodland songster begins its summer's night song, which is composed of downward spirals of flute-like notes. It spends the summer in our region, nesting on damp ground or on very low branches, but in autumn, it migrates to spend the winter in South America. **Where found:** moist, dense, deciduous woodlands with a heavy understorey, streamside thickets and riparian woodlands; from southern Newfoundland and southern QC to New Jersey in summer.

103

Hermit Thrush

Catharus guttatus

Length: 17 cm
Wingspan: 29 cm

The hermit thrush's lovely song is one of the finest in the forest, and this bird has a habit of quickly flicking its reddish tail into the air and then slowly lowering it. • This thrush is common in our area in summer, and a few may overwinter. If you spot a thrush in winter, it will most likely be a hermit. Perhaps this tendency to not follow along with the rest of its kin is how it got its reclusive name. **Where found:** forests and edge habitats; throughout Atlantic Canada in summer.

Wood Thrush

Hylocichla mustelina

Length: 20 cm
Wingspan: 33 cm

The clear, flute-like, whistled song of the wood thrush is one of the most beautiful and characteristic melodies of the eastern deciduous forest. A split syrinx, or vocal organ, enables the wood thrush to sing 2 notes simultaneously and thus create harmonies and hauntingly ethereal songs that delight listeners. • Still common but on the decline, the wood thrush faces loss of habitat and other threats, both here and in its Central American wintering habitat. **Where found:** moist, mature and preferably undisturbed deciduous woodlands and mixed forests; southern St. Lawrence region, Gaspé Peninsula and NB.

American Robin

Turdus migratorius

Length: 25 cm
Wingspan: 36 cm

The American robin is a familiar and common sight on lawns as it searches for worms. In winter, it switches to fruit trees, which can attract flocks to feed. • American robins build cup-shaped nests of grass, moss and mud. The female incubates 4 light blue eggs and raises up to 3 broods per year. The male cares for the fledglings from the first brood while the female incubates the second clutch of eggs. **Where found:** residential lawns and gardens, pastures, urban parks, broken forests, bogs and river shorelines; winters near fruit-bearing trees and springs; throughout Atlantic Canada in summer; remains in far eastern ranges year-round.

Gray Catbird

Dumetella carolinensis

Length: 22 cm
Wingspan: 28 cm

A gray catbird in full song issues a nonstop, squeaky barrage of warbling notes interspersed with poor imitations of other birds' songs. Occasionally it lets go with loud, cat-like meows that might even fool a feline. • The female catbird is one of the few birds that can recognize and remove a brown-headed cowbird egg sneakily laid in her nest. **Where found:** dense thickets, brambles, shrubby areas and hedgerows, often near water; throughout the Maritimes, southern QC and Newfoundland in summer.

Northern Mockingbird

Mimus polyglottos

Length: 25 cm
Wingspan: 35 cm

Masters of mimicry, mockingbirds can have a vocal repertoire of over 400 different song types. They imitate a wide array of sounds flawlessly, rivalling other birds at singing their own songs, mocking crows, and surprising and confusing humans with wolf-whistles, fire engine sirens and the backup beeps of garbage trucks. Male mockingbirds calling out for a mate have been known to sing through the night, to the frustration of anyone trying to sleep. **Where found:** hedges, fencerows and suburban parks and gardens; year-round in southern NS and NB; throughout the Maritimes in summer; sporadic breeder in all provinces.

Cedar Waxwing

Bombycilla cedrorum

Length: 18 cm
Wingspan: 30 cm

With its black mask and slick hairdo, the cedar waxwing has a heroic look. • To court a mate, the gentlemanly male hops toward a female and offers her a berry. The female will accept the berry and hop away, then stop and hop back toward the male to offer him the berry in return. **Where found:** *Breeding:* hardwood and mixed forests, woodland edges, fruit orchards, young pine plantations and among conifers in riparian hardwood stands. *In migration* and *winter:* open woodlands and brush, often near water; residential areas and any habitat with berry trees. In summer, throughout Atlantic Canada except Labrador; year-round in the very south of the region.

Ovenbird

Seiurus aurocapilla

Length: 15 cm
Wingspan: 24 cm

The ovenbird gets its name from the shape of its nest, which resembles an old-fashioned Dutch oven. • Finding an expertly concealed nest on the forest floor is nearly impossible, but these birds are easy to identify by sound. Issuing a loud *tea-CHER tea-CHER tea-CHER* song that ascends in volume, ovenbirds are conspicuous singers. • Unlike most other warblers, ovenbirds are primarily ground feeders and walk about poking through leaf litter for food. **Where found:** undisturbed, mature forests, often with little understorey; throughout Atlantic Canada in summer except northern QC and Labrador. **Also known as:** teacher bird.

Northern Waterthrush

Parkesia noveboracensis

Length: 15 cm
Wingspan: 24 cm

As its name suggests, this bird loves water, often wading in ponds and puddles, but it is a warbler, not a thrush. • This bird nests along the edges of shallow water and probes moist ground for insects and other small organisms. • The northern waterthrush bobs its tail constantly and rapidly as it walks along the ground in a similar manner to the ovenbird, to which it is closely related. **Where found:** riparian thickets, lakeshores and woodland bogs; throughout Atlantic Canada in summer. **Also known as:** *Seiurus noveboracensis.*

Black-and-White Warbler

Mniotilta varia

Length: 13 cm
Wingspan: 22 cm

In a habit unique to warblers but typical of nuthatches, the black-and-white warbler forages by creeping along branches and up and down tree trunks searching for insects in bark crevices. • This bird's song is reminiscent of a squeaky wheel, but it also has a dull *chip* call and a high-pitched flight note. **Where found:** mature deciduous and mixed woodlands; throughout Atlantic Canada in summer except northern QC and Labrador.

American Redstart

Setophaga ruticilla

Length: 13 cm
Wingspan: 22 cm

This bird's Latin American name, *candelita*,
meaning "little torch," perfectly describes the American
redstart. Not only are the male's bright orange patches the colour
of a glowing flame, but the bird never ceases to flicker, rhythmically
swaying its tail and flashing its orange wings, even when perched. • By flashing the
bright orange or yellow spots in its plumage, the redstart flushes insects from
the foliage. A broad bill and the rictal bristles around its mouth help it capture prey.
Where found: dense, shrubby understorey in deciduous woodlands, often near
water; throughout Atlantic Canada in summer except northern Labrador. **Also
known as:** butterfly bird.

Northern Parula

Setophaga americana

Length: 11 cm
Wingspan: 18 cm

This attractive warbler nests in our region, building
its nest inside a hanging mass of beard lichen. Its song
is a distinctive, high-pitched, rising buzz. • The northern
parula feeds on a large variety of insects and spiders; it will also eat fruit and visit
sugarwater and nectar feeders. **Where found:** coniferous and mixed woodlands
near water such as lakeshores, swamps and rivers; nests in coniferous or mixed
woods near water and lichens; throughout southern QC and the Maritimes in
summer. **Also known as:** *Parula americana.*

Yellow Warbler

Setophaga petechia

Length: 13 cm
Wingspan: 20 cm

Showy, bright yellow and common in
summer, the yellow warbler is a delight.
It is also a useful bird to have around
because it feeds on caterpillars, aphids and beetles.
• The yellow warbler is often parasitized by the brown-headed cowbird. It can
recognize cowbird eggs, but rather than tossing them out, it will build another
nest overtop the old eggs or abandon its nest completely. **Where found:** wetlands,
brushy fields, pond margins and scruffy woodland borders; throughout Atlantic
Canada in summer. **Also known as:** wild canary; *Dendroica petechia.*

Savannah Sparrow

Passerculus sandwichensis

Length: 14 cm
Wingspan: 20 cm

The savannah sparrow is inconspicuous in plumage and song, but it is one of our most common sparrows. You will typically observe it clinging to a swaying weed stalk. • This bird forages on the ground in search of seeds and insects but tends to fly directly to a raised perch if disturbed. **Where found:** in a wide variety of habitats, including grasslands, grassy beach dunes, farmlands and marshes; throughout Atlantic Canada in summer. **Also known as:** grey bird.

Song Sparrow

Melospiza melodia

Length: 15–18 cm
Wingspan: 20 cm

Although its plumage is unremarkable, the well-named song sparrow is among the great singers of the bird world. By the time a young male is only a few months old, he has already created a courtship tune of his own, having learned the basics of melody and rhythm from his father and rival males. • The presence of a well-stocked backyard feeder may be a fair trade for a sweet song in the dead of winter. **Where found:** hardwood brush in forests and open country, near water or in lush vegetation in riparian willows, marshy habitats and residential areas; throughout the southern half of the region in summer, absent from Labrador and northern Newfoundland.

White-throated Sparrow

Zonotrichia albicollis

Length: 17 cm
Wingspan: 22 cm

White-throated sparrows sing a distinctive, patriotic song—a clear, whistled *I love Canada Canada Canada*—in a somewhat mournful minor key. • White-throats have 2 colour morphs. One has black and white stripes on the head, whereas the other has brown and tan stripes. These colour morphs should not be misinterpreted as a difference between males and females. • This sparrow's diet consists mainly of insects, but it will feed on seeds if necessary, particularly in winter. **Where found:** coniferous and mixed forests; throughout Atlantic Canada in summer; year-round in southern NS. **Also known as:** Canada bird.

Dark-eyed Junco

Junco hyemalis

Length: 15–18 cm
Wingspan: 23 cm

Juncos usually congregate in sheltering conifers and at backyard bird feeders—with such amenities at their disposal, more and more are appearing in urban areas. These birds spend most of their time on the ground snatching up seeds underneath feeders, and are readily flushed from wooded trails. • There are 5 closely related dark-eyed junco subspecies in North America that differ in colouration and range. The slate-coloured junco occurs in our area. **Where found:** shrubby woodland borders and backyard feeders; throughout Atlantic Canada in summer; year-round in the Maritimes and southern Newfoundland. **Also known as:** black snowbird.

Northern Cardinal

Cardinalis cardinalis

Length: 22 cm
Wingspan: 30 cm

This historically southern bird has continued to expand its range northward over the last century. • The species is named for the colour of its plumage, which matches the red robes of Roman Catholic cardinals. • Cardinals maintain strong pair bonds. Some couples sing to each other year-round, whereas others join loose flocks, re-establishing pair bonds in spring during a "courtship feeding"—the male offers a seed to the female, which she then accepts and eats. **Where found:** woodland edges, thickets, backyards and parks; southern NB and NS year-round.

Snow Bunting

Plectrophenax nivalis

Length: 17 cm
Wingspan: 30 cm

The snow bunting is an Arctic species that would typically only spend winter in our region, but many of these birds are seen year-round. • With some populations living on the tundra where there are no trees, this bird has a habit of foraging on the ground for seeds and insects. It also nests in rocky cavities on the ground. **Where found:** tundra, rocky shores, beaches, grasslands and agricultural fields; throughout Atlantic Canada in winter except Labrador.

Red-winged Blackbird

Agelaius phoeniceus

Length: 19–23 cm
Wingspan: 33 cm

The male red-winged blackbird wears his bright red shoulders like armour—together with his short, raspy song, they are key in defending his territory from rivals. Nearly every cattail marsh worthy of note in our region hosts this bird and resonates with the male's proud and distinctive song. • The female's cryptic colouration allows her to sit inconspicuously on her nest, blending in perfectly among the cattails or shoreline bushes. **Where found:** cattail marshes, wet meadows and ditches, croplands and shoreline shrubs; southern half of the region in summer.

Common Grackle

Quiscalus quiscula

Length: 28–34 cm
Wingspan: 43 cm

The common grackle is a species of blackbird. It is not popular among birders because of its aggressive bullying at feeders or with farmers thanks to its habit of pulling up new corn shoots. • The male grackle's iridescent plumage gleams in sunlight. • Grackles form enormous roosts in autumn and winter with other blackbirds and European starlings (*Sturnus vulgaris*). Some of these roosts number hundreds of thousands of birds. **Where found:** nearly all habitats, especially in open to semi-open areas; throughout all but northernmost Atlantic Canada in summer; year-round in southernmost Atlantic Canada.

Brown-headed Cowbird

Molothrus ater

Length: 15–20 cm
Wingspan: 30 cm

Cowbirds are nest parasites and can be a serious problem for rare songbirds. They lay their eggs in other birds' nests and are known to parasitize more than 140 bird species. This habit evolved with the birds' association with nomadic bison herds—cowbirds were not in one place long enough to tend their own nests. Upon hatching, baby cowbirds outcompete the host's young, leading to nest failure. **Where found:** nearly ubiquitous; agricultural and residential areas, and woodland edges; throughout Atlantic Canada in summer except Labrador and northern Newfoundland; year-round in NS.

Purple Finch

Carpodacus purpureus

Length: 13–15 cm
Wingspan: 25 cm

Male purple finches are more raspberry red than purple. They are often confused with house finches (*C. mexicanus*), but the latter is more reddish and is prominently streaked below. Female purple finches have a bolder eyeline and facial pattern than do female house finches. • Purple finches are attracted to sunflower seeds, and large numbers can be lured to feeders in winter. **Where found:** coniferous and mixed forests; throughout Atlantic Canada in summer except northern QC and Labrador; becoming a more common breeder northward; year-round in southern regions.

White-winged Crossbill

Loxia leucoptera

Length: 15 cm
Wingspan: 24 cm

The bill of the white-winged crossbill seems like it should be a handicap, but, in fact, the crossed mandibles are adapted to pry open cones. This bird eats the seeds of spruce, fir and tamarack, and its bill is so well adapted for extracting seeds from cones that a single bird can eat up to 3000 conifer seeds per day! • When white-winged crossbills overwinter, they gather in flocks at the tops of spruce trees, creating showers of conifer cones and a crackling chatter of bills. **Where found:** primarily coniferous forests; throughout Atlantic Canada year-round.

Common Redpoll

Acanthis flammea

Length: 13 cm
Wingspan: 23 cm

These tiny snowploughs sometimes flock in the hundreds, gleaning bare fields for waste grain or stocking up at winter feeders. A high intake of food and the insulating layer of warm air trapped by their fluffed feathers keep these songbirds warm in cold climates. Redpolls can endure lower temperatures than any other songbird. **Where found:** weedy fields, roadsides, farmyards with spilled grain, backyards with feeders and woodland groves of seed-producing trees such as birch and alder; winters throughout Atlantic Canada; year-round only in northern NL.

Pine Siskin

Spinus pinus

Length: 11–14 cm
Wingspan: 23 cm

Pine siskins may be abundant for a time, then suddenly disappear. Because their favoured habitats are widely scattered, flocks are constantly on the move, searching forests for the most lucrative seed crops. • This sparrow-like bird has a characteristic, rising *zzzreeeee* call and chatters boisterously. **Where found:** coniferous forests, deciduous thickets, riparian areas and backyard feeders with black niger seed; from eastern QC to the western half of Newfoundland and the Maritimes in summer; year-round in southern areas. **Also known as:** *Carduelis pinus.*

American Goldfinch

Spinus tristis

Length: 11–13 cm
Wingspan: 23 cm

Like vibrant rays of sunshine, American goldfinches cheerily flutter over weedy fields, gardens and along roadsides, perching on late-summer thistle heads or poking through dandelion patches in search of seeds. It is hard to miss their jubilant *po-ta-to-chip* call and distinctive, undulating flight style. **Where found:** weedy fields, woodland edges, meadows, riparian areas, parks and gardens; year-round in the Maritimes; in summer also on the north coast of QC; absent from Labrador and northern Newfoundland. **Also known as:** willow goldfinch; *Carduelis tristis.*

House Sparrow

Passer domesticus

Length: 14–17 cm
Wingspan: 24 cm

This abundant and conspicuous bird was introduced to North America in the 1850s as part of a plan to control insects that were damaging grain and cereal crops. But, as it turns out, the house sparrow is largely vegetarian! • This bird will usurp the territory and nests of other native birds such as bluebirds, swallows and finches, and it has a high reproductive output of 4 clutches per year, with up to 8 young per clutch. **Where found:** townsites, urban and suburban areas, farmyards and agricultural areas, railway yards and other developed areas; throughout the region except northern Atlantic Canada.

AMPHIBIANS & REPTILES

Amphibians and reptiles are commonly referred to as "cold blooded," but this term is misleading. Although these animals lack the ability to generate their own internal body heat, they are not necessarily cold blooded. Amphibians and reptiles are ectothermic, or poikilothermic, meaning that the temperature of the surrounding environment governs their body temperature. These animals obtain heat from sunlight, warm rocks and logs, and warmed earth. In cold regions, reptiles and amphibians hibernate through winter, and some reptile species aestivate (are dormant during hot or dry periods) in summer in hot regions. Both reptiles and amphibians moult (shed their skins) as they grow.

Amphibians are smooth skinned and most live in moist habitats. They are represented by the salamanders, frogs and toads. These species typically lay eggs without shells in jelly-like masses in water. The eggs hatch into gilled larvae (the larvae of frogs and toads are called tadpoles), which later metamorphose into adults with lungs and legs. Amphibians can regenerate their skin and sometimes even entire limbs. Males and females often differ in size and colour, and males may have other specialized features when sexually mature, such as the vocal sacs present in many frogs and toads.

Reptiles are completely terrestrial vertebrates with scaly skin. In this guide, the representatives are turtles and snakes. Most reptiles bury their eggs in loose soil, but some snakes give birth to live young. Reptiles do not have a larval stage.

Salamanders	Toads & Frogs	Turtles	Snakes
pp. 114–116	pp. 116–118	pp. 118–119	pp. 120–121

Red-spotted Newt

Notophthalmus viridescens

Length: *Adult:* 7–14 cm; *Eft:* 4–8 cm

Like all amphibians, the newt morphs through several life stages. The larva hatches from an egg (the female lays up to 400 singly on submerged vegetation) after 1 to 2 months, but at the end of summer, rather than morphing into an adult newt, the larva develops into a terrestrial, nonbreeding eft, a stage that can last up to 4 years. The eft seeks shelter under logs and in rock crevices, but when it wanders through moist leaf litter, its vibrant red skin announces its toxicity to predators. When the eft returns to the water to morph into a mature breeding adult, a stage that can last another decade, its red skin dulls to green or brown with just a few red spots. **Where found:** moist forests; throughout Atlantic Canada. **Also known as:** eastern newt.

Blue-spotted Salamander

Ambystoma laterale

Length: 7–13 cm

The best time to observe this salamander is in spring when it emerges from underground, sometimes in the hundreds, to breed in ponds and wetlands. It is rarely seen again until late summer, particularly on rainy evenings, when newly transformed young emerge from their breeding ponds and migrate to terrestrial habitats. During these pilgrimages, the blue-spotted salamander deters hungry predators by secreting a foul-tasting liquid from glands at the base of its tail. **Where found:** mixed woodlands and continuous forests; may inhabit wetlands or lakes during breeding; throughout Atlantic Canada.

Spotted Salamander

Ambystoma maculatum

Length: 15–25 cm

Look into the depths of clear, shallow ponds that are void of fish in spring, and you may see this salamander walking about the bottom in search of a mate. The bright yellow spots make identification easy. You may also see several clumps of up to 250 eggs or a single, large mass of eggs laid by the female spotted salamander attached to underwater vegetation. The remainder of the adult's life cycle is spent under debris and loose soil in the forest. • This salamander feeds on invertebrates. **Where found:** forested areas surrounding ponds; Gaspé Peninsula, NS and NB.

Northern Two-lined Salamander

Eurycea bislineata

Length: 6–12 cm

The 2 lines that give this salamander its name are made up of many closely adjacent black spots, but a more distinctive identifying mark is the bold, yellow colou-ration between the 2 lines that runs the entire length of the animal—half of which is tail. • This salamander's long tail helps it navigate the currents of the fast-flowing streams along which it lives. **Where found:** cool, moist, forested areas alongside fast-moving streams; Gaspé Peninsula, coastal QC, Labrador and NB.

Four-toed Salamander

Hemidactylium scutatum

Length: 5–10 cm

If you find yourself wandering through a bog in summer, you may be one of the fortunate few to stumble across the little-known four-toed salamander. This amphibian loves to wander along the endless tunnels and passageways in sphagnum bogs, where few other animals, including humans, tend to roam. It is a terrestrial salamander, laying its eggs on land, where the larvae hatch and wiggle their way into small puddles and pools of water. • Invertebrates such as beetles, moths, spiders and mites are the preferred prey of this carnivore. **Where found:** sphagnum bogs, floodplains and along woodland streams; NS and NB.

Red-backed Salamander

Plethodon cinereus

Length: 6–13 cm

This type of salamander is lungless, meaning that instead of breathing with lungs, it breaths through its thin, moist skin. It belongs to a family of salamanders called Plethodontidae, which means "many teeth." • The red-backed salamander is abundant in almost any wooded environment, including woodlots, making it probably the most frequently encountered salamander in eastern Canada. Turn over a piece of rotting wood on your next summer hike, and you will likely see this small, slender, worm-like amphibian. **Where found:** damp, mature, deciduous or mixed forests, beneath coarse, woody ground litter and fallen, rotting logs; Gaspé Peninsula, NS, NB and PEI.

Spring Salamander

Gyrinophilus porphyriticus

Length: 11–21 cm

The bright red colouration of this lovely salamander is second only to that of the red-spotted newt. As with the newt, the colour is solely to deter predators—the spring salamander is colour blind and cannot admire this beauty in itself or its mate. • If you see this salamander, consider yourself lucky—and perhaps let a wildlife agency know where you observed it—because the spring salamander is one of Canada's rarest amphibians, with population estimates at fewer than 1000. **Where found:** cool, clear, well-oxygenated streams in forested mountainous areas; southeastern coastal QC.

American Toad

Anaxyrus americanus

Length: 6–11 cm

Touching a toad will not give you warts, but the American toad does have a way of discouraging unwanted affection—when handled, it may urinate on you! Despite this danger, children like to catch this large, reddish olive or brown toad. A more passive way of enjoying this amphibian is listening to its distinctive, long and loud croaking in spring. **Where found:** fields, woodlands, gardens and lawns; throughout much of Atlantic Canada but absent from northern NS, most of NB and most islands other than PEI.

Northern Spring Peeper

Pseudacris crucifer

Length: up to 3 cm

A grand chorus erupts from these tiny voices announcing the arrival of spring. Spring peepers will repeat their high-pitched peep up to 4000 times per hour! They sing most zealously from late afternoon through the night. These little frogs do not limit their operatic season to springtime but will continue throughout summer and autumn, until the snow flies. **Where found:** almost any shrubby or forested natural area near water; southern Gulf of St. Lawrence, NB, PEI and NS.

Wood Frog

Lithobates sylvatica

Length: 5–8 cm

A truly Canadian frog, this hardy animal is famous in the amphibian world for being able to survive being completely frozen! The frog's body produces complex sugars and proteins to draw water out of cells to prevent them from rupturing when frozen. The animal hibernates underground through winter, thaws out in spring and takes to still, icy ponds in search of heart-warming romance. **Where found:** moist woodlands, most still water bodies, tundra and grasslands; throughout most of Atlantic Canada except some northern islands, northern QC and northern Labrador; successfully introduced to Newfoundland. **Also known as:** *Rana sylvatica*.

Northern Leopard Frog

Lithobates pipiens

Length: 4–10 cm

Once the most widespread and abundant frog in Canada, this species suffered dramatic declines in the 1970s that it still has not recovered from in parts of its range. Much of the decline was attributed to water pollution from pesticides and herbicides, but it is unknown why this species suffered more than others. • This frog survives winter by remaining submerged in water below the ice. **Where found:** damp forests and wetlands; throughout the southern half of Atlantic Canada but also found in Labrador. **Also known as:** *Rana pipiens*.

Green Frog

Lithobates clamitans

Length: 7–10 cm

This is the frog you are most likely to catch anywhere in eastern Canada, though a big, fat individual could easily be mistaken for a bullfrog. Unlike the deep, croaking *jug-o-rum* of the bullfrog, however, the green frog's call is best described as a banjo-like twang—together with a bit of clog dancing it makes for a rhythmic Acadian summer night. **Where found:** near ponds, lakes and other permanent water bodies; throughout the Maritimes and along the St. Lawrence R.; introduced to Newfoundland. **Also known as:** *Rana clamitans*.

Bullfrog

Lithobates catesbeiana

Length: 15 cm or longer

Bullfrogs are very large, weighing from 30 to 850 grams or more, and long-lived, with an average lifespan of 7 to 9 years. There are records of individuals in captivity living up to 16 years. • Bullfrogs are predatory and will eat anything they can swallow, including other species of frogs, certain snakes, fish and even small birds and mammals. • The bullfrog's call is described as a deep, monotone *jug-o-rum more rum* (or *ouaouaron* in French). **Where found:** warm, still, shallow, vegetated waters of lakes, ponds, rivers and bogs; NB and NS. **Also known as:** *Rana catesbeiana.*

Red-eared Slider

Trachemys scripta

Length: 13–29 cm

Red-eared sliders, native to the southern U.S., were once common pets in Canadian homes, and some that either escaped or were set free (after the owners grew tired of them or the turtle outgrew its home) have survived in the wild and now compete with native turtles. They also transmit diseases and parasites to wildlife. The sale of red-eared sliders has been banned in many provinces, and organizations exist to find new homes for those whose owners no longer want them. **Where found:** wetlands; milder regions of the Maritimes and QC.

Painted Turtle

Chrysemys picta

Length: 10–25 cm

The bright yellow belly, red markings on the edges of the shell (carapace) and the distinct yellow stripes on the head are excellent diagnostics for identifying this striking turtle. • This reptile supplements its mainly vegetarian diet with invertebrates, amphibian larvae and small fish. **Where found:** shallow, muddy-bottomed water bodies with abundant plant growth; logs, cattail mats, sloughs, ponds, lakes, marshes and open banks along sluggish rivers with oxbows; southern QC, NB and NS.

Wood Turtle

Glyptemys insculpta

Length: 16–25 cm

Despite its sage looks and demeanour,
this mini-tortoise wants to be a clog dancer.
It stomps its feet on the ground to encourage earth-
worms to surface so it can eat them. • Although considered a sprinter among Canadian turtles, its top speed of
about 300 metres per hour is not fast enough to outrun poachers; this species has
been overharvested by collectors in the past, which, coupled with habitat loss,
has caused a decline in its population. **Where found:** woods, forests and wetlands;
hibernates in water; NB and NS. **Also known as:** *Clemmys insculpta*.

Common Snapping Turtle

Chelydra serpentine

Length: 20–47 cm

The common snapping turtle is the largest freshwater turtle in Canada. • These
turtles spend most of their existence wal-
lowing in the mud. The female only comes on
land in late spring to dig a nest in which to
lay her 25 to 50 eggs. Snapping turtles do not mate until
at least 15 years of age. • This species' reputation for biting
off toes is greatly exaggerated, but they do snap, so be wary.
Where found: soft, muddy bottoms of ponds and brackish water
with plenty of vegetation; from NS south to the Gulf of Mexico.

Leatherback

Dermochelys choriacea

Length: 1.3–1.4 m

Found in all the oceans of the
world, this sea turtle is not common, but it may enter our waters

in search of jellyfish. Typically, younger individuals are observed.
• Unlike other sea turtles with hard-plated carapaces, the leatherback has 7 smooth, leathery-looking ridges running the length of
the shell. • Loggerheads (*Caretta caretta*) and green sea turtles (*Chelonia mydas*)
may also be seen on rare occasions in our waters. **Where found:** offshore;
throughout Atlantic Canada.

Eastern Garter Snake

Thamnophis sirtalis

Length: 85–100 cm

Swift both on land and in water, the eastern garter snake is an efficient hunter of amphibians, fish, small mammals, slugs and leeches. It is a harmless and well-known Canadian snake, famous not only for its large hibernacula, but also for individuals hibernating in the basements of houses. • Several subspecies exist across the country with variable colouration; the Maritime subspecies is brown or grey, with a poorly defined spinal stripe and alternating dark marks on the back. **Where found:** wetlands, forests, fields and urban areas; throughout Atlantic Canada except NL; absent on some islands.

Red-bellied Snake

Storeria occipitomaculata

Length: up to 40 cm

This diminutive snake is not much larger than the worms it feeds upon. Its preferred prey species are the worms and slugs that it finds under logs and other forest debris in the woodlands in which it lives. • When angered or feeding, this snake curls its upper lip to expose its fangs. • The red-bellied snake is named for the colour of its belly, which can be red, orange or yellow. The rest of its body is brown, grey or black, with one to several dark or light dorsal stripes. **Where found:** open woodlands, forest edges and meadows, typically under boards, logs or other debris; NB, NS and PEI.

Smooth Green Snake

Liochlorophis vernalis

Length: 30–60 cm

This small snake is a master of disguise, as slender and green as a blade of grass. It also hangs from trees, swaying in the breeze much like a thin, green branch. It is attracted to the heat of road surfaces, where it is often run over by cars. If you find a small, blue snake dead on the side of the road, it is likely a green snake, which, oddly, changes colour when it dies. **Where found:** grassy areas, wet meadows and open woodlands; NB, NS and PEI.

Northern Ring-necked Snake

Diadophis punctatus

Length: 25–60 cm (average less than 46 cm)

Despite its outstanding good looks, this snake is shy and prefers to hide beneath rocks, logs, bark, wooden planks or leaf debris. If it is discovered and threatened, this harmless, pencil-thin snake will show its brightly coloured underside by coiling its tail upward like a fiery corkscrew, hide its head beneath its body and emit a pungent musk. • This snake's preferred prey is the red-backed salamander. **Where found:** wooded areas and nearby meadows; NB and NS.

FISH

Fish are ectothermic vertebrates that live in the water, have streamlined bodies covered in scales and possess fins and gills. A fundamental feature of fish is the serially repeated set of vertebrae and segmented muscles that allow the animal to move from side to side, propelling it through the water. A varying number of fins, depending on the species, further aid the fish to swim and navigate. Most fish are oviparous and lay eggs that are fertilized externally. Spawning is an intense time for fish, often involving extraordinary risks. Eggs are either produced in vast quantities and scattered, or they are laid in a spawning nest (called a "redd") under rocks or logs. All these methods are designed to keep the eggs healthy and surrounded by clean, oxygen-rich water. Parental care may be present in the defence of such a nest or territory.

Trout, Salmon & Whitefish
pp. 122–123

Smelt
p. 123

Pike
p. 123

Suckers
p. 124

Minnows
p. 124

Killifish
p. 124

Sculpins
p. 125

Eels & Lampreys
p. 125

Sturgeon
p. 126

Cod
p. 126

Flounders
p. 127

Mackerels & Tunas
p. 127

Alewife
p. 128

Skates
p. 128

Sharks
p. 128

Rainbow Trout

Oncorhynchus mykiss

Length: 19–46 cm

This Pacific native was introduced to lakes for angling and accidently to marine environments through aquaculture. The latter form of this trout is called the steelhead, and it has an anadromous life cycle similar to that of salmon, meaning that it spawns in freshwater but spends the remainder of its life in the ocean. • When spawning, this trout shows the colours of its namesake, with a greenish to bluish back, silvery sides and belly that are often tinged with yellow and green and a reddish lateral line. **Where found:** lakes and coastal rivers; from the Avalon Peninsula in Newfoundland, across the southern Maritime provinces from NS to ON. **Also known as:** steelhead.

Atlantic Salmon

Salmo salar

Length: up to 1.5 m

Instinct draws the salmon back to its freshwater spawning grounds after up to 4 years in the ocean. After surpassing all obstacles, including predators and an arduous upstream swim, the female salmon lays her eggs and the male fertilizes them before they both expire in the shallow waters of their birthplace. Only the fittest survive to return to the ocean. • Some populations are landlocked. **Where found:** open sea close to coasts; deep, cold lakes if landlocked; spawns in clear, oxygen-rich, flowing waters; from northern QC south. **Also known as:** summer or fall salmon; parr, smolt, grilse and kelt during specific life stages.

Brook Trout

Salvelinus fontinalis

Length: up to 20 cm

Colourful and feisty, the brook trout is a type of char and is one of the most sought-after game fish. It is native to Atlantic Canada and has a freshwater, or resident, form as well as a sea-run, or anadromous, form. The resident form is smaller and darker than the sea-run. • Unlike salmon, sea-run "brookies" stay near the coast, moving in and out of river estuaries. During spawning runs, they race up their home rivers to the headwaters, lay and fertilize their eggs, and then return to the sea. **Where found:** lakes, coastlines, freshwater streams and river estuaries; throughout Atlantic Canada. **Also known as:** brookie, brook char, speckled trout, coaster trout, sea trout.

Lake Whitefish

Coregonus clupeaformis

Length: 40–50 cm

The lake whitefish is what biologists call a "plastic species," which
means that the species changes its behaviour, food habits and appearance in different habitats. One of the best identifiers for the different forms is the number of
gillrakers. Fish that live in more open water develop extended gillrakers that are
better for filtering plankton. Lake whitefish caught closer to the surface tend to
have higher gillraker counts than those that nibble food from the lake bottom.
Where found: cool, deep water at the bottom of larger lakes; occasionally in rivers;
QC and NL. **Also known as:** humpback whitefish, eastern whitefish.

Rainbow Smelt

Osmerus mordax

Length: up to 20 cm

The commercial smelt industry
in Canada originated on the Atlantic coast, but
the rainbow smelt was one of the first fish to be experimented with as a gillnet
fishery on the Great Lakes, starting in 1948, where it flourished. Sport fishing for
smelt is also popular, as is ice fishing on several small New Brunswick lakes where
there are landlocked smelt populations. **Where found:** cool, deep, offshore waters
as well as lakes; from Hamilton Inlet, Labrador, south along the coast; landlocked
populations are found in various lakes throughout the region. **Also known as:**
American smelt.

Chain Pickerel

Esox niger

Length: 60 cm (large males may exceed
this length)

There are a few species of pike in Atlantic Canada, and where you are determines
which one you will encounter. The chain pickerel occurs from Nova Scotia south
all the way to Florida, but if you are in Labrador, you will see northern pike
(*E. lucius*), and in the St. Lawrence region, you will find the famed muskellunge,
or "muskie" (*E. masquinongy*), and, where their ranges overlap, a hybrid of the
muskie and the northern pike called the tiger muskie occurs. All the species are
similar in appearance and have noteworthy long, large snouts full of razor-sharp
canine and conical teeth. **Where found:** warm, clear, vegetated lakes and slow-
flowing rivers and streams. **Also known as:** jackfish, pike.

White Sucker

Catostomus commersoni

Length: 25–60 cm

This generalist species lives in habitats ranging from cold streams to warm, even polluted, waters. It avoids rapid currents and feeds in shallow areas. • During the spring spawning season, mating white suckers splash and jostle in streams or along shallow lakeshores. Suckers migrate upstream shortly after the ice breaks up, providing an important food for other fish, eagles and bears. Once hatched, the fry provide critical food for other young fish. **Where found:** varied habitats; prefers shallow lakes and rivers with sandy or gravel substrate; tolerant of turbidity and stagnant water; throughout Atlantic Canada. **Also known as:** common sucker, mud sucker, brook sucker.

Emerald Shiner

Notropis atherinoides

Length: 5–13 cm

This baitfish is abundant in larger rivers and lakes and is important to many predators, both aquatic and avian. Its populations fluctuate greatly, influencing the populations of many other fish species in the process. • These minnows spend much of their time in open water feeding on plankton, which they follow up to the surface at dusk. In autumn, large schools of these little gems gather near shorelines and docks. **Where found:** open water of lakes, large rivers and shallow lakeshores in spring and autumn; throughout Atlantic Canada. **Also known as:** lake shiner, common shiner, buckeye shiner.

Banded Killifish

Fundulus diaphanus

Length: 5–13 cm

These small, schooling fish provide important food for mudpuppies (*Necturus* spp.), birds and larger fish, including bass and pike. They are named for the vertical stripes or bands that run along their bodies. • Killifish are members of the topminnow family, so named because these species typically forage at or near the water's surface, where they can easily be observed. **Where found:** shallows of warmer lakes and slow-moving rivers; from Newfoundland south, including the St. Lawrence R. basin.

Mottled Sculpin

Cottus bairdii

Length: 8–15 cm

The sculpin is famous for its
looks—it is so ugly that it is beautiful.
Bulging eyes, fat, wide lips, rough-textured skin with mottled colouration and
several dorsal spines add up to one visually impressive fish. • When stressed, this
fish expands its gills and emits a low-pitched, humming sound. **Where found:**
shallow coastal waters with sandy bottoms; spawns in bays, estuaries and saltwater sloughs; some adults move to deep water in non-spawning seasons;
throughout Atlantic Canada.

American Eel

Anguilla rostrata

Length: *Male:* seldom exceeds 40 cm;
Female: may exceed 1 m

Eels are catadromous fishes—when
they reach maturity, they migrate from
fresh water to the ocean to spawn. However, some
eels stay in coastal or estuarine waters until maturation or migrate periodically
between river and estuary. • The environment influences eel gender—crowding or
poor conditions result in more males; in larger rivers, most eels become female.
For example, the St. Lawrence River has almost exclusively female eels, whereas
the poorly productive East River, in Chester, Nova Scotia, has approximately
60 percent male eels. **Where found:** south from Newfoundland and the Gulf of
St. Lawrence along the Atlantic coast to the Gulf of Mexico; northern continental
limit is the Hamilton Inlet–Lake Melville Estuary of Labrador.

Sea Lamprey

Petromyzon marinus

Length: to 90 cm

Lampreys look eel-like, but they
are not eels. Lampreys lack jaws
and instead have a prominent sucking disc filled with large, hooked teeth. They also
have several gill openings and lack pectoral fins, and instead of a backbone or
cartilage, they have a primitive pliable notochord. They are parasitic, latching onto
host fish to feed on blood and tissue. Strangely, some adults are non-parasitic and
will go without food until they spawn and die. • Lamprey larvae live in silt and mud
on the ocean floor for up to 5 years. **Where found:** rivers, estuaries, along coasts
and in open waters; from the Gulf of St. Lawrence south to northern Florida.

Atlantic Sturgeon

Acipenser oxyrinchus oxyrinchus

Length: to 5.5 m

Sturgeons are one of the oldest fish species in the world, and individuals are themselves long-lived, reaching up to 60 years of age, by which time they can weigh around 360 kg. They are coveted for their caviar—it is the European sturgeon or giant beluga (*Huso huso*) of the Adriatic, Black and Caspian seas that is most prized for this delicacy, leading to poaching of this now-endangered species. • Distinctive features of the Atlantic sturgeon are the 5 scutes (large, bony shields) and long nose. Sturgeons lack teeth because they are bottom feeders, sucking up their dinner from the ocean floor like giant vacuums. **Where found:** rivers and ocean; from the Maritimes south. **Also known as:** sea sturgeon.

Burbot

Lota lota

Length: 12–40 in

The burbot is the only member of the cod family confined to fresh water. • The single chin barbel and the pectoral fins contain taste buds. As these fish grow, they satisfy their ravenous appetite for whitefish and suckers by eating larger fish instead of increased numbers of smaller ones, sometimes swallowing fish almost as big as themselves. • Once considered by anglers to be a "trash" fish, the burbot is gaining popularity among sport fishers. **Where found:** bottom of cold lakes and rivers; throughout Atlantic Canada. **Also known as:** freshwater cod, eelpout, ling, lawyer, loche.

Atlantic Cod

Gadus morhua

Length: 1.8 m

Stories will ever abound about the Atlantic cod, which was a critical species that sustained the Atlantic marine ecosystem and whose great schools provided livelihoods for the Atlantic fisher folk. Whether we ever see the return of such numbers is doubtful, and the trickle effects of impacts to wild species that rely on the Atlantic cod for food are yet to be seen. • The maximum mature size of this fish is another thing perhaps lost—once recorded at weights up to 90 kg, today this fish is rarely seen at half that size, most commonly 2.7 to 5.4 kg. **Where found:** in deep waters on or near the bottom of the continental shelf; from western Greenland south to North Carolina, but most abundant (historically) from Labrador to New York.

Winter Flounder

Pseudopleuronectes americanus

Length: 64 cm

Masters of disguise, flounders lie flat and
partially buried in sand on the ocean floor and change
their pigmentation and patterns to exactly match the
sand and pebbles around them. • Young flounders actu-
ally look like normal fish, but the left eye slowly moves to the right (in some species
it is the right eye that moves to the left), and the fish remains on one side of its
body at rest and while swimming for the rest of its adult life. **Where found:** sandy
bottoms in shallow coastal waters in winter; from NL south to Georgia. **Also
known as:** blackback flounder.

Atlantic Mackerel

Scomber scombrus

Length: up to 56 cm

Travelling in large schools, this fish
is migratory in habit and is an abundant and important species for commercial
fisheries. Often harvested as a baitfish or feed for fish farms, this small fish is more
sustainable than larger fish species for feeding fish-hungry nations, and it is higher
in omega fatty acids and lower in bio-accumulated toxins. • The chub mackerel
(*S. japonicus*) is another of the 23 North American mackerel species and is present
from Nova Scotia to Florida. **Where found:** open waters in temperate regions; from
Newfoundland south.

Atlantic Bluefin Tuna

Thunnus thynnus thynnus

Length: up to 3 m

The Atlantic bluefin tuna is warm
blooded, which is rare among fish, and
can therefore maintain its body temperature in cold water and during
deep dives (to 1000 m). This allows the bluefin to range into Canadian waters during
summer and autumn. • Since sushi became popular outside Japan, the global
demand for this premium fish has led to overharvesting and to concerns for the
sustainability of this fishery and, more importantly, for the survival of this species.
• The tuna's natural predators include sharks and whales. • The bluefin is a large
species and can weigh up to 450 kg. **Where found:** throughout the Atlantic Ocean.

Alewife

Alosa pseudoharengus

Length: less than 30 cm

The alewife and a closely related species, the blueback herring (*A. aestivalis*), are commonly referred to interchangeably as gaspereau in Atlantic Canada and as river herring along the Atlantic coast of the United States. The term "alewife" can represent either species because both have a somewhat similar appearance and biology. Historically, they were used as baitfish, but now they are commercially valuable. • Both species are anadromous, returning from the sea to natal rivers to spawn. **Where found:** rivers and the ocean; Newfoundland and the Gulf of St. Lawrence south; abundant in the Miramichi, Margree, LaHave, Tusket, Shubenacadie and Saint John rivers. **Also known as:** gaspereau (Canada), river herring (U.S.).

Thorny Skate

Amblyraja radiata

Length: just over 1 m

Skates and rays are species of shark but are more reminiscent of birds as they gently flap their "wings" while swimming in their watery sky. Skates stay close to the ocean floor, searching for crustaceans, fish and shellfish prey, and keep their eyes looking skyward for predators such as larger sharks. • Leathery "mermaid purses," 10 to 15 cm long, are skate egg cases, in which 1 to 2 baby skates develop. • The thorny skate is one of the most common skates found in the Grand Banks, the Gulf of St. Lawrence, the Scotian Shelf, the Bay of Fundy and Georges Bank. **Where found:** over hard and soft bottoms, mainly offshore, at depths of 18–1400 m; from western Greenland, Davis Strait, Hudson Strait, Hudson Bay and off Labrador south as far as South Carolina.

Porbeagle Shark

Lamna nasus

Length: 3.7 m

The porbeagle shark is potentially dangerous to humans because of its size (230 kg), but conflicts are extremely rare; this big fish prefers to feed on smaller fish such as herring, mackerel, flounder and cod, as well as squid and other sharks. • The porbeagle is an endangered species, but it is still commercially and recreationally hunted in the Atlantic and caught as bycatch by commercial fisheries. • Porbeagle sharks prefer deep, cold water and have the ability to thermoregulate, which is rare in fish, to maintain body temperature. **Where found:** continental shelf and open water but does occur inshore; from Newfoundland to New Jersey.

INVERTEBRATES

More than 95 percent of all animal species are invertebrates, and there are thousands of invertebrate species in our region. The few mentioned in this guide are frequently encountered and easily recognizable. Invertebrates can be found in a variety of habitats and are an important part of most ecosystems. They provide food for birds, amphibians, shrews, bats and other insects, and they also play an important role in the pollination of plants as well as aiding in the decay process.

Barnacles
p. 133

Bivalves
pp. 133–134

Sea Snails
pp. 134–135

Sea Urchins & Sea Stars
p. 135

Sea Anemones
p. 136

Crustaceans
pp. 136–137

Jellyfish
p. 137

Squid
p. 137

Butterflies & Moths
pp. 138–139

Dragonflies
p. 140

Beetles
pp. 140–141

Grasshoppers
p. 141

Wasps & Bees
pp. 141–142

Two-winged Flies
p. 142

Harvestmen
p. 142

Acorn Barnacle

Balanus spp.

Diameter: to 10 cm

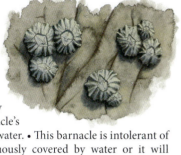

We typically see this barnacle closed, but when it feeds, though rarely and sometimes not for months at a time, long, feathery plumes reach out from the top of the barnacle's shell to filter bits of organic matter from the water. • This barnacle is intolerant of exposure and must remain almost continuously covered by water or it will become desiccated. • Capable of sexual reproduction, yet immobile, this animal has the largest penis-to-body size ratio of all animals, necessary so that it can reach its mate. **Where found:** rocky shores and exposed coasts; lower intertidal zone with continuous water cover; subtidal zone to depths of 90 m.

Softshell Clam

Mya arenaria

Length: 5–7 cm

Frequently the clam of choice at a clambake, this bivalve is also a major food source for many shore-birds, invertebrates and fish. Beyond its role as a meal, it also serves a valuable function in filtering and cleaning water sources, and does so to such an effective degree that it is often used as a water-quality control tool by agencies enforcing pollution standards. • This clam's white to pale grey shell varies in colour depending on the minerals in the sand into which it burrows. **Where found:** brackish waters and estuaries; burrows to depths of 30 cm; typically in upper to lower intertidal zones but also in deep water to 190 m. **Where found:** from Labrador to Cape Hatteras. **Also known as:** eastern softshell clam, long-necked clam, sand gaper, nannynose, steamer clam; *M. hemphillii*.

Blue Mussel

Mytilus edulis

Length: 3–10 cm

Great colonies of blue mussels can be seen covering rocks, wooden pilings or anything solid and stationary to which they can anchor themselves. • In clean waters, these mussels are popular edible shellfish. Blue mussels are common and widespread, found also on the west coast and even in Europe, where they are commercially farmed and are a much more commonly eaten mussel than in North America, though their popularity here is increasing. **Where found:** near the low tide line; along the coast from the Arctic to South Carolina. **Also known as:** edible mussel.

American Oyster

Crassostrea virginica

Length: 5–25 cm

More than half of the oysters harvested in the U.S. and Canada are American oysters, a common species along the eastern seaboard and particularly important as a commercial fishery and a popular culinary delicacy. • This shellfish has prolific reproductive abilities—a female spawns 10 to 20 million eggs on average, and a large female may spawn up to 100 million. • An oyster changes gender throughout its life cycle, with larger, older individuals remaining female. **Where found:** on hard or soft substrates in low-salinity waters on the ocean floor at depths of 3–12 m; from the Gulf of St. Lawrence to Florida and the Gulf of Mexico. **Also known as:** eastern oyster.

Atlantic Deep-sea Scallop

Placopecten magellanicus

Diameter: to 20 cm

Because it inhabits deep waters (to 75 m, though as shallow as 2 m in estuaries and bays), you are unlikely to see this scallop in the wild, but you will likely find it in fish markets and restaurants. You might also find its almost circular, somewhat two-toned shell (with the upper, rounded valve ranging from yellow to purplish and the lower, flat valve being white) washed up on the beach. **Where found:** sandy or gravelly ocean floors at depths of 4–122 m; from Labrador to North Carolina. **Also known as:** giant scallop, ocean scallop, smooth scallop.

Northern Moon Snail

Euspira heros

Length: 11 cm

This snail's shell is named after the moon, and a full one at that—the shell is quite spherical, almost as wide (89 mm) as it is long, and pale grey to white in colour. There are typically 5 convex whorls. • When the animal is disturbed, it retracts into its shell and closes a sort of trap door, called an operculum. • If you find a little round fort, or collar, of sand, you will have discovered the "egg collar" where this snail has laid its eggs. **Where found:** sandy beaches and ocean bottoms at the low tide line to depths of 366 m or greater; from Labrador to North Carolina. **Also known as:** northern moon shell.

Common Periwinkle

Littorina littorea

Length: 25 mm

Hundreds of these little snails can be seen grazing on algae on the surface of rocks, seaweed or even the shells of other molluscs. • A periwinkle can remain out of the water for long periods of time without becoming desiccated by closing off the entrance to its shell with a tightly fitting trap door called an operculum. • This little snail was introduced from Europe. **Where found:** on rocks, other shellfish or seaweed at low to high tide lines; from Labrador to Maryland. **Also known as:** rough periwinkle, smooth periwinkle.

Green Sea Urchin

Strongylocentrotus droebachiensis

Diameter: 83 mm

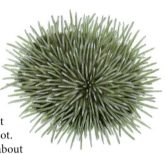

Be careful when wading in the calm waters of protected bays—you risk stepping on this abundant sea urchin, a very painful experience that may leave broken spines imbedded in your foot. • The sea urchin's round shell, called a test, is about 38 mm tall in the centre, with the spines being a maximum of about a third of this height. When the animal dies, the spines fall off, and the empty test can be found on shore; the bumps on the shell are where the spines once attached. **Where found:** rocky shores and kelp beds, from low tide to depths of 1157 m; from the Arctic to New Jersey.

Daisy Brittle Star

Ophiopholis aculeate

Width: to 10 cm

This common sea star is usually reddish in colour, but can sometimes have dark markings or variable colouration. • The daisy brittle star feeds mainly on bivalves, wrapping its 5 to 7 long (up to 8 cm) arms around them and forcing them out of their shells. Its own predators include other sea stars, molluscs and crustaceans. The brittle star's escape strategy is to detach whichever arm may be in the predator's grasp or even in curious human hands, hence the name "brittle star." **Where found:** hidden within or beneath rocks in tidal pools, burrowed in the sand and mud; also in lower intertidal zones; from NL south to Cape Cod.

Frilled Anemone

Metridium senile

Height: 30 cm
Width (crown): 2–5 cm

Ornately beautiful when its multitude of tentacles dance in the currents, this anemone looks like no more than a blob of jelly when the tentacles are retracted. • If this animal finds itself out of water at low tide, it will cover itself with a layer of sand. • In deep, subtidal waters this anemone can grow to 50 cm tall and 25 cm wide. • The frilled anemone is typically white, but other pale colours, including yellow, pink, orange, red, grey, brown and olive green, occur. **Where found:** adheres to rocks, artificial structures and shells along sandy, muddy or rocky shorelines where there are medium to strong currents, from the low tide line to depths of 165 m; all along the Atlantic shoreline.

Green Crab

Carcinus maenas

Width: 79 mm

Introduced from Europe, this small scavenger is probably the most common little seashore crab you will encounter out of the several species that can be seen scampering across rocks and hiding in tidal pools. • This crab has been accidentally introduced to our shores and to others around the world by stowing away on ships. It has flourished in its new habitats and become a pest, forcing local shellfish and other organisms out of house and home. **Where found:** in tidal pools, mud banks and wetlands, from open shores to brackish water; from Newfoundland south. **Also known as:** European crab.

Northern Shrimp

Pandalus borealis

Length: maximum 17 cm

Commercial trawling for this deep-water shrimp at depths of 20 to 1330 m, usually on soft, muddy bottoms, feeds the demand for this crustacean. It is served in restaurants as "prawns" and sold in most large supermarkets. • This shrimp's red colour camouflages it in deep water, where red wavelengths of light cannot reach, thus rendering the northern shrimp nearly invisible. **Where found:** deep ocean waters; from the Arctic to Cape Cod. **Also known as:** Maine shrimp, boreal shrimp, red shrimp, pink shrimp, deep-water prawn, deep-sea prawn.

North American Lobster

Homarus americanus

Length: 86 cm

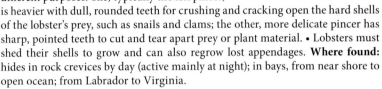

The lobster turns red when cooked, but it is otherwise greenish, or rarely yellow or blue. • The 2 pincers serve different purposes: one, typically the left pincer, is heavier with dull, rounded teeth for crushing and cracking open the hard shells of the lobster's prey, such as snails and clams; the other, more delicate pincer has sharp, pointed teeth to cut and tear apart prey or plant material. • Lobsters must shed their shells to grow and can also regrow lost appendages. **Where found:** hides in rock crevices by day (active mainly at night); in bays, from near shore to open ocean; from Labrador to Virginia.

Moon Jellyfish

Aurelia aurita

Diameter: 40 cm

This ethereal whitish to translucent medusa is a favourite food of the leatherback sea turtle but not a favoured acquaintance of swimmers and snorkelers—the animal can deliver a painful sting, and it also releases polyps in the water that are difficult to see but easily felt. The sting may cause a slight rash or itching for several hours. • The moon jellyfish has 8 lobes fringed by numerous short tentacles and 4 long, oral arms with frilly margins. **Where found:** floats near the water's surface just offshore; often washes up on beaches during high tide or after a storm; throughout the Maritimes. **Also known as:** moon jelly.

Short-finned Squid

Illex illecebrosus

Length: 18–31 cm

These squid are caught and used more often as fish bait than for calamari. They are used to catch commercially important fish such as sea bass and Atlantic mackerel, which are the main predators of squid. • The squid avoids predators by being a fast swimmer (it can propel itself more than 30 km/h by shooting water from its body), camouflaging the hue of its skin or by releasing clouds of dark ink to distract pursuers as it flees from sight. **Where found:** in large schools near the water's surface and to depths of 91 m over the continental shelf; from the Bay of Fundy south. **Also known as:** boreal squid.

Cabbage White

Pieris rapae

Wingspan: 50 mm

This diminutive butterfly flits about the garden looking pretty, but its caterpillar is the bane of gardeners, who find holes bored in their vegetables by the hungry juvenile. • Butterflies need heat to fuel their wings, which is why they are often seen sunning themselves with their wings spread. The dark bases of this butterfly's wings help absorb solar heat and transfer it to the flight muscles, while the white wing surfaces reflect light inward to heat the butterfly's body. **Where found:** any open habitat, especially gardens and agricultural fields; avoids dense forest; throughout Atlantic Canada.

Eastern Black Swallowtail

Papilio polyxenes

Wingspan: 8–22 cm

This gorgeous black butterfly is, lucky for us, one of the most common to frequent our backyard gardens. • The variably coloured caterpillar, which is typically white and green with black bands and yellow spots, has an orange osmeterium (a type of gland) that looks like a forked snake tongue. The organ not only deters predators upon sight, but also omits a foul odour if the caterpillar is threatened. **Where found:** open fields, meadows, backyards and roadsides; throughout Atlantic Canada.

Mourning Cloak

Nymphalis antiopa

Wingspan: 70 mm

This hibernating butterfly usually sleeps away the winter under a piece of bark or the siding of a barn, in a woodpile or even in a window shutter, but if the temperatures are above freezing, you may see it even in winter. The mourning cloak emerges at the first hint of spring and is usually the first butterfly to be seen, sometimes as early as April. • This butterfly breeds in its second year of adulthood, having metamorphosed the previous July. This yearlong adulthood is a record in the butterfly world. **Where found:** openings in forested areas; throughout Atlantic Canada.

American Copper

Lycaena phlaeas

Wingspan: 25 mm

There are dozens of copper species across Canada. Their colouration is a combination of orange and brown, and sometimes purple, giving a coppery lustre to their wings. • These butterflies attempt to be inconspicuous and stay low to the ground, but they do appear in residential gardens to sip from the flowers. In good years, they can produce two generations, the second being more abundant. **Where found:** open or disturbed sites, pastures, landfills and roadsides; from the Maritimes south.

Spring Azure

Celastrina ladon

Wingspan: 25 mm

This dainty blue azure endears itself to us by being one of the first butterflies to announce the arrival of spring. It feeds on the buds and flowers of spring blooms. An adult lives only for only a week or two, in which time it must breed and lay its eggs. The larvae often develop on the leaves of dogwood and cherry trees, and may be tended to and protected by ants for the sweet honeydew that they produce. • Spring azures belong to a family of butterflies known as "blues" because of their upperwing colouration, but only the males sport the colour; the females are dull brown. **Where found:** forest clearings; throughout Atlantic Canada.

Luna Moth

Actias spp.

Wingspan: about 95 mm

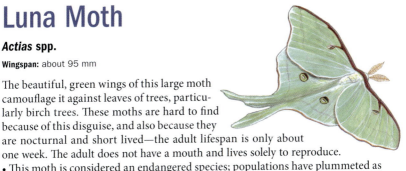

The beautiful, green wings of this large moth camouflage it against leaves of trees, particularly birch trees. These moths are hard to find because of this disguise, and also because they are nocturnal and short lived—the adult lifespan is only about one week. The adult does not have a mouth and lives solely to reproduce. • This moth is considered an endangered species; populations have plummeted as a result of pollution and pesticide use. **Where found:** deciduous forests; southern Atlantic Canada.

Green Darner

Anax junius

Length: up to 80 mm

This gorgeous dragonfly is an agile and effective hunter, and one that we like to see in abundance because it feasts on our most irritating insects, mosquitoes and blackflies. The aquatic larvae also do their share of pest control, devouring the larvae of mosquitoes and other insects. • The male darner is mostly blue with a green thorax; the female is green with a grey or brown abdomen. • This migratory insect spends the summer in our area. **Where found:** near ponds and lakes; throughout Atlantic Canada.

Common Whitetail

Plathemis lydia

Length: 50 mm

There are well over one hundred dragonfly and damselfly species in Altantic Canada, many of which are hard to distinguish from each other; however, the common whitetail is unmistakeable. The checkered wings and the male's robust, chalky blue-white body (females are shorter and brown) makes him an eye-catching dragonfly. • The feminine scientific name is a mystery—no one knows who Lydia was, but she was likely the muse of an ancient entymologist. **Where found:** perched or hawking for mosquitoes above the surface of ponds, marshes and slow-moving rivers; throughout Atlantic Canada except at higher altitudes; absent from Labrador. **Also known as:** long-tailed skimmer.

Firefly

Photuris pennsylvanica

Length: about 12 mm

At night, fireflies seem like fairies dancing in the forest. They light up in a beautiful, sparkling courtship dance, using the time-honoured strategy of attracting a mate with something shiny and sparkly! • The insects can control the chemical reaction that causes their abdomens to glow a bright yellow-green. The larvae also glow and are known as "glow worms." • Fireflies belong to the beetle family and hatch from eggs as larvae. **Where found:** deciduous forests; throughout southern Atlantic Canada.

Multicoloured Asian Ladybug

Harmonia axyridis

Length: 5 mm

There are several ladybug species, all distinguishable from each other by size, number of spots and colouration. Some are not even red. • This voracious species was introduced to the U.S. as early as 1916 as a biological control against crop damage because it feeds ravenously on aphids; it has since spread throughout North America, outcompeting many of our native ladybugs. **Where found:** open areas, hilltops and urban gardens in spring and autumn; throughout Atlantic Canada.

Migratory Grasshopper

Melanoplus sanguinipes

Length: to 30 mm

As per its name, this grasshopper is a traveller, dispersing widely and occasionally swarming, but never to biblical proportions. At high densities, however, it is a serious pest, causing devastating damage to crops. It will feed on wheat, barley, oats, alfalfa, clover, corn, garden vegetables and ornamentals, including vines, bushes, fruit trees and bark. • This grasshopper is inactive in the cool of night, but after basking in the sun, it begins jumping and flying about. **Where found:** grasslands and meadows; also agricultural, suburban and urban areas; from Newfoundland south.

Eastern Yellow Jacket

Vespula maculifrons

Length: 10–15 mm

These predators kill a variety of other insects and sometimes feed on nectar at flowers. They are attracted to sweets such as fruit, juices and sodas. • Big and boldly striped in black and yellow, these ill-tempered hornets nest in ground burrows, often those made by small mammals. Yellow jackets also build amazing paper nests that are often even more elaborate than those of the true paper wasps (*Polistes* spp.). **Where found:** nearly ubiquitous in open areas; throughout Atlantic Canada. **Also known as:** hornet.

Bumble Bee

Bombus spp.

Length: 7–20 mm

Bumble bees are found mainly in northern temperate regions and range much farther north than honey bees (*Apis* spp.); colonies can be found in the Arctic on Ellesmere Island. However, from the entire colony, only the queen survives the winter. • Although bumble bees are not usually aggressive, they will sting if harmed or to defend their nest. Unlike a honey bee, a bumble bee's stinger lacks barbs, so it can sting more than once. **Where found:** clearings and meadows wherever there are flowering plants; throughout Atlantic Canada.

Horsefly

Tabanus americanus

Length: 20–28 mm

Horseflies are known for their nasty bite, which can bleed for several minutes because of anticoagulants in the flies' saliva. They are strongly associated with wet habitats, where they lay eggs in masses attached to plants overhanging fresh water. When the larvae hatch, they fall into the water and overwinter in the muddy bottom. **Where found:** swamps, salt marshes and ponds; throughout Atlantic Canada.

Eastern Daddy Longlegs

Leiobunum spp.

Length: 5–8 mm

Daddy longlegs have tiny bodies and very long, thin legs up to 40 times longer than the body. These arachnids are not actually spiders, which are in the order Araneae, they are harvestmen and belong to the order Opiliones. Daddy longlegs do not make webs, and they feed on decaying vegetable and animal matter. **Where found:** open areas on foliage, tree trunks and shady walls on buildings; throughout Atlantic Canada. **Also known as:** harvestman.

PLANTS

Plants belong to the Kingdom Plantae. They are autotrophic, which means they produce their own food from inorganic materials through a process called photosynthesis. Plants are the basis of all food webs. They supply oxygen to the atmosphere, modify climate, and create and hold soil in place. Plants disperse their seeds and pollen through carriers such as wind, water and animals. Fossil fuels come from ancient deposits of organic matter—largely plants. In this book, plants are separated into 3 categories: trees; shrubs and vines; and herbs, grasses, ferns and seaweeds.

sugar kelp

ostrich fern

red oak

swamp rose

TREES

Trees are long-lived, woody plants that are normally taller than 5 metres. There are 2 types of trees: coniferous and broadleaf. Conifers, or cone-bearers, have needles or small, scale-like leaves. Most conifers are evergreens, but some, such as larches (*Larix* spp.), shed their leaves in winter. Most broadleaf trees lose their leaves in autumn and are often called deciduous trees (deciduous means "falling off" in Latin). Some exceptions include rhododendrons and several hollies.

A single tree can provide a home or a food source for many different animals. Tree roots bind soil and play host to a multitude of beneficial fungi, and even support certain semi-parasitic plants such as Indian-pipe (*Monotropa uniflora*). Trunks provide a substrate for numerous species of mosses and lichens, which in turn are used by many animals for shelter and nesting material. Tree cavities are used by everything from owls to squirrels to snakes. Leafy canopies support an amazing diversity of life. Myriad birds depend on mature trees, as do scores of insects. Both the seed cones of coniferous trees and the fruits of deciduous trees are consumed by all manner of wildlife.

A group of trees can provide a windbreak, camouflage or shelter, and can hold down soil, thus preventing erosion. Streamside (riparian) woodlands are vital to protecting water quality. Their dense root layers filter out sediments and other contaminants that would otherwise enter watercourses. It is no mystery why the healthiest rivers are those buffered by abundant, undisturbed woodlands. There are many types of forest communities, and their species composition is largely dictated by the type of soil on which they occur. To a large extent, the types of trees within a forest determine what other species of plants and animals are present. Old-growth forest is critical habitat for many species that use the fallen or hollowed-out trees as nesting or denning sites. Many species of invertebrates live within or under the bark, providing food for birds. Fallen, decomposing logs provide habitat for snakes, salamanders, mosses, fungi and invertebrates. The logs eventually completely degrade into nutrient-rich soil to perpetuate the continued growth of plant life and keep organic matter in the ecosystem. Large forests retain carbon dioxide, an important preventive factor of global warming. One giant old-growth tree can extract 7 kg of airborne pollutants annually and put back 14 kg of oxygen. Responsibly managed forests can also sustain an industry that provides wood products and jobs.

Tree heights provided in the following accounts reflect the usual size and shape of each species, but trees can reach greater heights in ideal conditions. Conifers have both pollen cones (male) and seed cones (female), but only mature seed cones are described. They are useful in identifying conifers because they are found year-round on the tree or on the ground nearby. Leaf, cone and fruit measurements and descriptions are given for mature plants only.

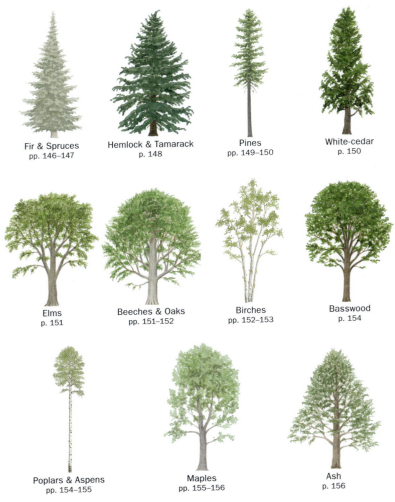

Fir & Spruces
pp. 146–147

Hemlock & Tamarack
p. 148

Pines
pp. 149–150

White-cedar
p. 150

Elms
p. 151

Beeches & Oaks
pp. 151–152

Birches
pp. 152–153

Basswood
p. 154

Poplars & Aspens
pp. 154–155

Maples
pp. 155–156

Ash
p. 156

Balsam Fir

Abies balsamea

Height: to 25 m
Needles: 2–2.5 cm long, dark green, flat, blunt, flexible, aromatic
Seed cones: 5–10 cm long, erect, barrel-shaped, greyish brown

Balsam fir is the provincial tree of New Brunswick and has a beautiful fragrance, making it also a popular Christmas tree—it fills the home with that traditional pine scent, with the added benefit of not dropping its needles soon after being cut. • A balsam fir can be identified by the flat, blunt (not pointed) needles, which have 2 white lines on the lower surface, and the cylindrical cones that usually grow near the top of the tree's spire-like crown. The scales fall from the cones in autumn leaving "candles," unlike other conifers, which drop the entire cone. • The oil-rich seeds are eaten by porcupines and many species of birds and squirrels. **Where found:** from low, swampy ground to well-drained hillsides; moist soils, cut areas and fields; from sea level to timberline; from Labrador south.

White Spruce

Picea glauca

Height: 25 m; old-growth to 43 m
Needles: 13–20 mm long, stiff, 4-sided, green, often with a greyish bloom
Seed cones: 5–8 cm long, cylindrical, pale brown

White spruce can live for 200 years and eventually replaces pines in mature forests in a process called succession. This tree is a good choice for landscaping and is used in reforestation. • Spruce needles roll between your fingers, unlike the flat, 2-sided needles of fir. • Traditionally, Native peoples used flexible spruce roots to lace together birchbark canoes and resin to waterproof the seams and holes. • White spruce seeds are eaten by many birds and mammals, and the low-hanging lower boughs, when covered in snow, create shelter for grouse, rabbits and other wildlife. **Where found:** various soils and climates, but prefers moist, rich soil; from sea level to timberline; from Labrador south.

Black Spruce

Picea mariana

Height: to 15 m; rarely to 30 m
Needles: 8–15 mm long, stiff, 4-sided
Seed cones: 15–40 mm long, oval becoming round when open, dull greyish brown to purplish brown

This slow-growing wetland tree, which may live for 200 years, is an important source of lumber and pulp. • Northern explorers used this tree to make spruce beer, a popular drink that prevented scurvy, as well as spruce gum or syrup to treat coughs and sore throats. • Snowshoe hares love to eat young black spruce seedlings, and red squirrels harvest the cones, but in general, this tree is not favoured by wildlife as a food source. • Black spruce is the provincial tree of Newfoundland and Labrador. **Where found:** moist habitats, cool, boggy sites, swamps and coastal areas; at elevations of 610–1524 m; common throughout Atlantic Canada from Labrador south.

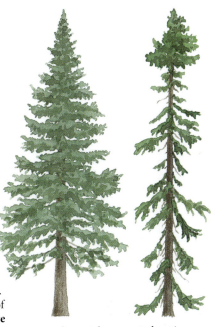

Red Spruce

Picea rubens

Height: 15–24 m
Needles: 1–2.5 cm long, shiny, yellow-green, curved, 4-sided
Seed cones: 2–4 cm long, oval, brown or reddish brown when mature, hanging at branch tips

The needles of red spruce give a bottlebrush look to the stems of this tree. • The cones hang near the top of the tree and drop in autumn, so are not present in winter. • The needles are yellowish green, but the dark brown bark has a reddish hue, giving this spruce its common name. It is a beautiful tree and has been adopted as the provincial tree of Nova Scotia. **Where found:** prefers open, well-drained sites; growth is stunted in wet habitats; east to NS and south to New England, at elevations to 1372–1981 m.

Eastern Hemlock

Tsuga canadensis

Height: 20–30 m; occasionally to 45 m
Needles: less than 2 cm long, flat, flexible, blunt, unequal, in 2 opposite rows
Seed cones: less than 2 cm long, brown and dry, hanging at ends of twigs

These trees can live for 600 years, with some living for a millennium; there were once pure stands of eastern hemlocks, with trees reaching 1 metre in diameter. Some of these ancient trees can still be found in old-growth Acadian forests. • The brittle wood separates along annual rings, making it easy to split, but it often "pops" and sparks when burned. The wood also has hard knots and frequently twists, but it remains a popular lumber for outdoor products such as lawn furniture because it is rot resistant. • Eastern hemlock provides an abundant food supply and dense cover for many animals, including white-tailed deer, snowshoe hares, seed-eating birds and ruffed grouse. **Where found:** cool, sandy, moist sites, ravines and riparian areas; found east to Cape Breton I. and south to New England, to elevations of 914 m in the north and elevations of 610–1524 m in the south.

Tamarack

Larix laricina

Height: 12–24 m
Needles: 2–2.5 cm long, soft, deciduous, tightly spiralled in tufts of 15–60
Seed cones: 12–19 mm long, upright, rounded, rose red turning to pale brown

The tamarack is unique among our conifers in that its needles turn golden yellow and drop in autumn, making this tree a deciduous conifer. Tamaracks are referred to as the "yellow candles" of autumn woodlands. • The straight, oily, moisture-resistant trunks are used as poles, piers and railway ties. The tannin-rich bark was used for tanning leather. The sap contains a natural sugar gelatin that tastes like bitter honey. **Where found:** open sites, old fields and moist, well-drained sites; also bogs, swamps and muskeg; from Labrador south, from near sea level to 518–1219 m in the southern part of its range. **Also known as:** larch, hackmatack.

Eastern White Pine

Pinus strobus

Height: to over 30 m; formerly to 46 m or more
Needles: 5–15 cm long, slender, flexible, soft, 3-sided, in bundles of 5
Seed cones: 8–20 cm long, brown, woody, cylindrical

The airy, graceful white pine is the largest North American conifer and once occurred in vast virgin stands that were estimated to contain 2 billion m^3 of lumber. When European settlers arrived, the trees were rapidly cut down to make everything from ship masts to matchsticks. • This valuable tree provides food and shelter for many species of wildlife, including chickadees, red squirrels and porcupines. • The scientific name *strobus* means "gum-yielding" or "pitchy tree" in Latin and "cone" in Greek. **Where found:** from dry sand, rocky ridges and gravelly soils to sphagnum bogs and humid sites with well-drained soil; from southeastern MB east to Newfoundland and south, at elevations from near sea level to 610 m.

Jack Pine

Pinus banksiana

Height: 9–21 m; potential maximum 27 m
Needles: 2–4 cm long, flattened, stiff, sharp-pointed, twisted, in bundles of 2
Seed cones: 3–5 cm long, yellowish brown, woody, can be closed, straight or curved inward

Jack pines are the first conifers to colonize clear-cuts and areas burned by fire. The cones are held shut with a tight resin bond that melts when heated and allows the seeds to disperse. The seeds are all be released at the same time, which is why stands of jack pine are often made up of trees of the same age. • Cones usually occur in groups of 2 to 3 and point toward the tip of the branch. • Animals and birds browse on young jack pine seedlings and eat the fallen seeds. **Where found:** dry, infertile, acidic, often sandy or rocky soils; from NS south, at elevations to about 610 m.

Red Pine

Pinus resinosa

Height: 21–24 m; rarely over 30 m
Needles: 11–16.5 cm long, shiny, dark green, brittle, sharp-pointed, in bundles of 2
Seed cones: 4–6 cm long, oval, light brown, woody, hanging at upper branch tips

Red pines live about 200 years. Slow growing at first, once established, they can shoot up 30 cm per year. • This tree is valuable for its timber, which is easily preserved because of the high creosote content in the porous wood. It has been popular for building ship masts, poles and cribbing, as well as for other outdoor uses. The yellowish to reddish colour of the wood is not desirable for furniture. • The bark and seeds provide winter forage for birds and mammals. **Where found:** dry sand and gravelly, well-drained soils; east to NS and south; local in Newfoundland; at elevations of 213–427 m. **Also known as:** Norway pine.

Northern White-cedar

Thuja occidentalis

Height: 12–21 m
Needles: 3 mm long, dull green, scale-like, overlapping, gland-dotted
Seed cones: 7–12 mm long, upright, green becoming reddish brown with age

Fragrant cedar lumber is known for resisting decay, but living trees are frequently hollow from heart rot. The wood is often used near water in cedar-strip canoes, shingles and dock posts. • This species can live for a long time. A magnificent specimen on the Niagara Escarpment is at least 1050 years old. It is commonly named arborvitae, which means "tree of life." • Native peoples and French settlers used parts of the tree to prevent scurvy. Deer often browse the lower branches, and pine siskins are fond of the seeds. **Where found:** humid habitats; swamps and high snowfall areas, often in pure stands; east to NS and south to New England, to elevations of 914 m in southern parts of its range. **Also known as:** arborvitae.

American Elm

Ulmus americana

Height: to 30 m
Leaves: 7.5–15 cm long, alternate, oval, rounded and asymmetrical at base, pointed at tip, prominently veined
Flowers: 3 mm across, greenish, in tassel-like clusters
Fruit: samaras 10–12 mm long, 1-seeded

Large, graceful elm trees once lined our city streets and parks, but hundreds of thousands have been lost to Dutch elm disease since its arrival in the United States in 1930. This elm is abundant in the wild but seldom gets very large before being attacked and killed by the fungal infection. Occasional giants still occur as isolated specimens in the midst of agricultural fields, where the disease cannot readily reach them. • American elms are important as hosts for many species of moths and butterflies, such as mourning cloaks. **Where found:** valleys, floodplains, stream terraces, sheltered slopes and moist soils; also in mixed hardwood forests; east to Cape Breton I. and south to New England, to elevations of 762 m.

American Beech

Fagus grandifolia

Height: 18–24 m
Leaves: 6–13 cm long, alternate, narrowly oval, parallel veins each end in a coarse tooth
Flowers: male flowers 20–25 mm long in dense, hanging clusters; female flowers 6 mm long in small, erect clusters
Fruit: prickly burs 12–19 mm long, greenish to reddish brown, enclosing pairs of 3-angled, shiny, brown nuts

The edible nuts of mature beech trees are said to taste best after the first frost but should be eaten in moderation to avoid an upset stomach. Ground, roasted beechnuts were traditionally used as a coffee substitute. The oil was extracted and used as food and lamp oil. Beechnuts have a high fat content and are an important food for animals, including squirrels and black bears. • A tiny, parasitic wildflower called beechdrops (*Epifagus virginianus*) grows on the roots of beeches and is found on or near the host tree. **Where found:** moist, well-drained slopes and bottomlands; east to Cape Breton I. and south to New England, to elevations of 914 m.

Red Oak

Quercus rubra

Height: 18–27 m
Leaves: 10–23 cm long, alternate, dull yellowish green, deeply pinnately lobed, with 7–11 nearly triangular lobes
Flowers: tiny; male flowers in hanging catkins 10–12.5 cm long; female flowers in small clusters
Fruit: acorns 15–28 mm long, lower ¼ of nut seated in a saucer-shaped cup

This common eastern tree is the provincial tree of Prince Edward Island and a member of the red or black oak group, which includes black oak (*Q. velutina*) and pin oak (*Q. palustris*). These species all have deep, pointed leaf lobes, bitter acorns that ripen in 2 years and non-scaly bark. • Although the acorns contain tannins that are toxic to humans, they are an important food for squirrels, raccoons, black bears, white-tailed deer and birds. • The copious flowers of mature oaks attract masses of insects, which are consumed by birds travelling from the tropics to boreal nesting grounds. **Where found:** rich, moist, loamy, sandy or clay soils, often in pure stands on the lower and middle slopes of forests; east to Cape Breton I. and south to New England, to elevations of 1676 m.

Yellow Birch

Betula alleghaniensis

Height: 21–30 m
Leaves: 7.5–13 cm long, alternate, yellowish green, toothed, bright yellow in autumn
Flowers: tiny; male flowers in catkins 5–10 cm long; female flowers in erect, cone-like catkins 13–20 mm long
Fruit: small, flat, 2-winged nutlets, in hanging catkins 2–3 cm long

North American birches are divided into two groups: yellow birches, which include the yellow birch and cherry birch (*B. lenta*), and white birches, which include the white birch and grey birch (*B. populifolia*). Members of the yellow birch group have leaves with 8 to 12 straight veins per side, slender, cone-like catkins and inner bark that smells and tastes like wintergreen; white birches have leaves with fewer veins and inner bark without a distinct fragrance. • Yellow birch sap can be boiled into syrup or fermented into beer; the leaves and twigs can be used to make a fragrant tea. **Where found:** rich, moist, often shady sites; east to southern Newfoundland and south to New England, at elevations to 762 m.

White Birch

Betula papyrifera

Height: 15–21 m
Leaves: 5–10 cm long, alternate, dull green,
toothed, slender-pointed, 5–9 straight veins per side
Flowers: tiny; male flowers in hanging catkins 5–10 cm long,
in clusters of 1–3; female flowers in erect catkins 2–4 cm long
Fruit: small, flat, broadly 2-winged nutlets, in narrow, brown,
hanging catkins 4–5 cm long

The smooth, pale bark of this birch peels off in papery sheets and was used by Native peoples to make birchbark canoes, baskets and writing paper. Never peel the bark off a living tree because the tree can be scarred or killed. • Birch wood is suitable for making sleds, snowshoes, paddles, canoe ribs, arrows and tool handles. • Betulic acid, the compound that makes birch bark white, is being studied for use in sunscreens and in the treatment of skin cancer. **Where found:** dry to moist, open or disturbed sites (prefers full sun and nutrient-rich soils) and forest edges; east to Labrador and south to New England, at elevations to 1219 m. **Also known as:** paper birch, canoe birch.

Eastern Hop-hornbeam

Ostrya virginiana

Height: 6–15 m
Leaves: 5–13 cm long, alternate,
dark yellowish green, sharp-
toothed, pointed
Flowers: tiny; male flowers
greenish, in dense, cylindrical, hanging catkins 1.5–5 cm long; female flowers reddish
green, in small, loose catkins 5–8 cm long
Fruit: hop-like, hanging clusters of flat,
yellowish, papery sacs 1–2 cm long, each
sac with a single nutlet

Often planted as an ornamental tree, eastern hop-hornbeam is slow growing. • The dense, resilient wood is used to make tool handles, fence posts and other durable wood items. It is so hard that it is difficult to split with an axe. • The fruits resemble hops, an ingredient in beer, hence this tree's common name. • Ruffed grouse and squirrels are among the wildlife that feed on the buds and nutlets; white-tailed deer browse on the twigs. **Where found:** moist understorey in hardwood forests; east to Cape Breton I. and south to New England, at elevations to 1372 m. **Also known as:** ironwood.

American Basswood

Tilia americana

Height: 18–30 m
Leaves: 7.5–15 cm long, alternate, heart-shaped, sharply toothed, asymmetrical at base
Flowers: 12–15 mm across, 5 creamy yellowish petals, fragrant, in loose, hanging clusters
Fruit: brown, woolly, round, nut-like capsules 6–8 mm across, in long-stalked clusters

American basswood, with its fragrant flowers, large leaves and rounded crowns, is often planted in urban parks and gardens. It is the northernmost basswood species. • The soft wood is ideal for carving and is also used for furniture, measuring sticks and pulp. Linden flower tea, sold in health-food stores, provides a remedy for coughs, colds and bronchitis. • Bees attracted in droves to the hanging clusters of flowers produce strongly flavoured honey. **Where found:** cool, moist, rich woodlands, often near water and mixed with other hardwoods; east to southwestern NB and south to New England, at elevations to 975 m. **Also known as:** American linden, bee-tree.

Balsam Poplar

Populus balsamifera

Height: 18–24 m, twice that height possible
Leaves: 6–13 cm long, alternate, dark green above, silvery below, oval, blunt-toothed, tapered to a point
Flowers: tiny, male and female flowers on separate trees; male catkins 2–3 cm long; female catkins 8–20 cm long
Fruit: oval capsules, 6–8 mm across, numerous in hanging catkins 10–13 cm long; capsules release fluffy masses of tiny, brown seeds tipped with soft, white hairs

This tree is both the largest poplar species and the largest hardwood in general in North America. • In spring, many Native groups ate the young catkins and sweet inner bark. Medicinally, the leaves, bark and resinous aromatic buds were important for treating many conditions and ailments—the resins are still used today in salves, cough medicines and painkillers. • The wood is ideal for campfires, because it does not crackle and makes clean smoke. **Where found:** moist to wet, low-lying sites, often on shorelines; also foothills to subalpine zones; east to Newfoundland and south to New England, from sea level to mid elevations. **Also known as:** black cottonwood, balm-of-Gilead; *P. trichocarpa*.

Quaking Aspen

Populus tremuloides

Height: 12–21 m, sometimes taller
Leaves: 3–7.5 cm long, alternate, dark green above, paler beneath, oval with a short point, finely blunt-toothed
Flowers: tiny, in slender, hanging catkins 2.5–6 cm long, male and female flowers on separate trees
Fruit: light green, cone-shaped capsules, 5–7 mm long, in hanging catkins up to 10 cm long; capsules release many tiny, cotton-tipped seeds

The common names "quaking" and "trembling" refer to the way the leaves flutter in the slightest breeze because of the narrow leaf stalks; this trait is a good way to differentiate this tree from the similar balsam poplar. • Aspen trunks were once used for tipi poles and canoe paddles; today, aspen wood is harvested primarily for pulp and for making chopsticks. • Suckers from the shallow, spreading roots of this deciduous tree can colonize many hectares of land. Single trunks are short-lived, but a colony of clones can survive for thousands of years. **Where found:** dry to moist sites; east to Newfoundland and south to New England, from near sea level to elevations of 1982–3048 m in southern parts of its range. **Also known as:** trembling aspen, aspen poplar.

Sugar Maple

Acer saccharum

Height: 21–30 m
Leaves: 9–14 cm long and wide, opposite, 5 palmate lobes, irregularly coarse-toothed
Flowers: 5 mm across, yellowish green, on slender, hairy stalks in tassel-like clusters
Fruit: winged samaras 2.5–3 cm long, hanging in pairs

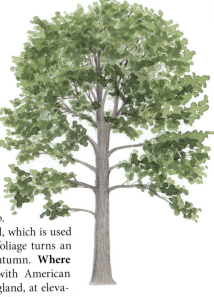

Sugar maples are famous worldwide for their sap, the main source of pure maple syrup and tasty maple sugar. Each spring, festivals celebrate the traditions of boiling sap and making maple syrup. About 40 L of sap yield 1 L of maple syrup. • Sugar maple is also prized for its wood, which is used in high-quality furniture. • This tree's foliage turns an exceptionally showy orange-red in autumn. **Where found:** deep, rich, moist soils, often with American beech; east to NS and south to New England, at elevations to 762 m. **Also known as:** hard maple.

Red Maple

Acer rubrum

Height: 18–27 m
Leaves: 6–10 cm long and wide, opposite,
palmately 3–5 lobed, irregularly double-toothed
Flowers: 3 mm across, reddish, in tassel-like clusters
Fruit: reddish or yellow, winged samaras 2–2.5 cm
long, hanging in pairs

The red twigs, buds, flowers and autumn leaves of our national emblem add colour to North America's eastern forests. The bright red flower clusters appear in early spring. • Red maple leaves have palmate lobes separated by shallow notches. The samara wings spread at a 50° to 60° angle. • The even, straight-grained wood is used in cabinets, furniture and flooring, and the bark can be boiled into a red ink or dark brown dye. The sap yields syrup that is semisweet. **Where found:** cool, moist sites near swamps, streams and springs; sometimes in drier upland sites; east to Newfoundland and south to New England, at elevations to 1829 m.

White Ash

Fraxinus americana

Height: to 24 m
Leaves: 20–30 cm long, opposite, dark green,
paler beneath, smooth margins, compound,
pinnately divided into 5–9 (usually 7) sharp-
pointed leaflets 6–13 cm long
Flowers: 6 mm across, purplish to yellow,
in compact clusters along twigs
Fruit: slender, winged samaras 2.5–5 cm
long, hanging in clusters

This tree is North America's main source of commercial ash. The strong, flexible wood is used in sporting goods (especially base-ball bats), tool handles, boats and church pews. Historically, plough, airplane and automobile frames were made from ash. • Native Americans made a yellow dye from the bark. • Ash leaves can be crushed and applied to skin to soothe mosquito and bee stings. **Where found:** among other hardwoods on upland sites with rocky to deep, well-drained soils; east to Cape Breton I. and south to New England, at elevations to 1524 m.

SHRUBS & VINES

The difference between a tree and a shrub is sometimes rather sketchy, but in general, shrubs are small, woody plants less than 6 metres tall, though many may grow to 10 m if conditions permit. They are typically bushy, with multiple small trunks and branches that emerge from near the ground, and many species produce soft berries. Some shrubs occur in open, sunny areas, whereas others are important dominant components of the forest understorey. Shrubs provide habitat and shelter for a variety of animals, and their berries, leaves and often bark are crucial sources of food. The tasty berries of some shrubs have been long been a staple of Native and traditional foods, and they are still enjoyed by people throughout our region.

Vines are climbing or trailing, woody-stemmed plants. The growth form of a vine is based on long stems, which enables the plant to quickly colonize large areas or to take advantage of small patches of fertile soil—the vine can root in the soil, but grow so that its leaves have access to areas with more light. Vines use different climbing methods—some twine their stems around a support, whereas others use tendrils. Most vines are flowering plants and can be woody, such as riverbank grape, or herbaceous, such as wild cucumber.

Willows
p. 159

Alders
p. 159

Witch-hazel
p. 159

Heaths
pp. 160–161

Currants & Roses
pp. 161–164

Dogwoods
p. 165

Winterberry
p. 165

Sumac & Poison Ivy
p. 166

Buttonbush
p. 166

Honeysuckles
p. 167

Bayberry
p. 167

Wild Cucumber
p. 168

Bittersweet
p. 168

Grape
p. 168

Pussy Willow

Salix discolor

Height: 2–6 m
Leaves: 3–10 cm long, oblong to elliptical, wavy-edged
Flowers: tiny, in hairy catkins 2–4 cm long
Fruit: long-beaked, hairy capsules 6–12 mm long

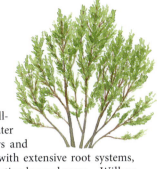

Pussy willow, with its gorgeous, furry catkins, is a well-known and well-loved species. • All willows are water loving and are most commonly found along rivers and streams, as well as on hillsides. Fast growing and with extensive root systems, willows are good for erosion control and for re-vegetating burned areas. • Willow branches are popular for wickerwork. Green branches can be used to smoke meat. **Where found:** moist to wet open sites; throughout Atlantic Canada.

Speckled Alder

Alnus incana

Height: to 10 m
Leaves: 4–10 cm long, alternate, thick, broadly ovate, toothed
Flowers: tiny, in dense catkins; male catkins are 5–8 cm long, slender, loose and hanging; female catkins are 1–2 cm long, woody and cone-like
Fruit: tiny, flat, 2-winged nutlets in the seed cones

This large, open shrub takes over open areas in early stages of forest succession—before much else takes hold, alders appear and make conditions more suitable for other plants. Their roots stabilize the soil as well as imparting nitrogen-fixing bacteria that improve soil quality. The leaves also contain nitrogen, which enters the soil when the leaves fall and decompose. **Where found:** dry hillsides and near coasts; throughout the Maritimes.

Witch-hazel

Hamamelis virginiana

Height: 6 m
Leaves: 6–15 cm long, alternate, oval, scalloped
Flowers: small, fragrant, bright yellow, 4 ribbon-like petals, clustered in 3s along twigs
Fruit: woody, light brown capsules 10–14 mm across

Well known for its astringent properties and ability to stop bleeding, witch-hazel preparations are widely available commercially for topical use. The leaves and bark of this shrub are used in medicinal extracts, skin cosmetics, shaving lotions, mouthwashes, eye lotion, ointments and soaps. • The edible, oily seeds seem to explode from their capsules, which fire the seeds up to 10 m away. **Where found:** moist, shaded areas and transitional forests; throughout the Maritimes. **Also known as:** winterbloom.

Labrador-tea

Rhododendron groenlandicum

Height: 50–150 cm
Leaves: 2–6 cm long, alternate, evergreen, oblong to elliptical, leathery, dark green and wrinkled above, rusty-woolly beneath, edges curled under
Flowers: 1 cm across, fragrant, white, in rounded clusters
Fruit: drooping, dry, brown capsules, 5–6 mm across

This small shrub's leaves are thick and leathery, with rolled edges and distinctive reddish, woolly hairs on their undersides, all adaptations to help conserve moisture. • Native peoples and early settlers made the leaves and flowers into a tea that was rich in vitamin C. However, consuming large amounts of this tea can be toxic. **Where found:** moist, acidic, nutrient-poor soils, coniferous forests and cool bogs; associated with black spruce; NL. **Also known as:** *Ledum groenlandicum.*

Bog Laurel

Kalmia polifolia

Height: usually less than 1 m
Leaves: 1–4 cm long, opposite, evergreen, elliptical, leathery, shiny and dark green above, whitish below, edges rolled under, on flattened stems
Flowers: small, bell-shaped, rose pink, 5-parted with petals connected almost to tips, in clusters of 2–5
Fruit: round, woody, 5-lobed capsules, 6 mm across

This small shrub's leathery leaves curl under and their undersides are covered with fine hairs to help prevent moisture loss; the leaves are tinged red in winter. • Bog laurel contains poisonous andromedotoxin compounds that can cause breathing problems, vomiting and even death if ingested in sufficient quantities. Even honey made from the nectar is said to be poisonous. • **Where found:** bogs; throughout the Maritimes. **Also known as:** bog rosemary, kalmia heath.

Leatherleaf

Chamaedaphne calyculata

Height: to 1.5 m
Leaves: 3–4 cm long, alternate, elliptical, leathery, scaly, dark green turning red-brown in winter
Flowers: 5–6 mm across, white, bell-shaped, in one-sided racemes
Fruit: round, brownish, many-seeded capsules, 3–5 mm across

In May, bogs come alive with the snowy flower racemes of this shrub, which often forms dense thickets or floating mats at the edges of open-water swamps. • Leatherleaf is a member of the heath family (Ericaceae). Heaths are important in the ecological makeup of bogs, helping to regulate and store water in the system. **Where found:** wet sphagnum peat bogs, lake margins, wet areas, barrens, seacoasts and boreal forest; common around bogs and lakes; throughout Atlantic Canada.

Highbush Blueberry

Vaccinium corymbosum

Height: 1.5–4.5 m
Leaves: 4–8 cm long, alternate, elliptical, dark green, red in autumn
Flowers: 6–13 mm across, greenish white to pinkish, urn-shaped, waxy, in clusters
Fruit: dull blue to black berries, 1 cm across

Plentiful blueberries were an important fruit for northern Native peoples, and blueberry picking remains a favourite tradition today. • There are many *Vaccinium* species native to the Maritimes, with many more that have been introduced and cultivated, including blueberries and cranberries. **Where found:** variety of habitats; reaches peak abundance in bogs, sometimes dry, sandy areas and open, upland woods; also transitional forests; from QC to NS and south.

Mayflower

Epigaea repens

Height: 10–15 cm, creeping, mat-forming
Leaves: 2–7.5 cm long, alternate, evergreen, leathery, oval
Flowers: 0.5–1.5 cm wide, trumpet-shaped, 5 lobes, white to pink, fragrant, in small clusters
Fruit: whitish, berry-like capsules, 4–5 mm across

The mayflower, the floral emblem of Nova Scotia, was once plentiful but is becoming scarcer because of its sensitivity to environmental disturbances. Stewards of pristine environments are rewarded with the exquisite fragrance of this lovely native plant, which trails a string of flowers along the forest floor. **Where found:** sandy, acidic soils in exposed sites in coniferous and mixedwood forests; from Newfoundland south. **Also known as:** trailing arbutus.

Wild Black Currant

Ribes americanum

Height: 1 m
Leaves: 3–8 cm long, slightly wider, alternate, 3–5 pointed lobes, coarsely double-toothed edges
Flowers: small, creamy white to yellowish, bell-shaped, in hanging clusters
Fruit: black berries, 6–10 mm across, in drooping clusters

Native peoples ate the fruit of this plant and other currant species. The young leaves were cooked and eaten with raw fat. Currants are high in pectin and make excellent jams and jellies. Mixed with other berries, they are used to flavour liqueurs or make wines. • Raw currants tend to be very tart, but these common shrubs provide a safe emergency food source. **Where found:** damp soil along streams, wooded slopes, open meadows and rocky ground; from QC to NB.

Pin Cherry

Prunus pensylvanica

Height: to 10 m
Leaves: 6–11 cm long, alternate, oval to lance-shaped
Flowers: 6–10 mm wide, white, 5 round petals, in flat-topped clusters
Fruit: bright red cherries, 4–8 mm wide

Unpopular raw because of their sourness and tiny pits, pin cherries are nonetheless a bountiful fruit that is delicious in pies, jelly and wine. Native peoples dried them and pounded them into animal fat and meat to make pemmican, which was once an important staple. • This shrub is one of the first trees to regenerate after a fire. **Where found:** moist, forested areas, clearings, burned areas and transitional forests; throughout Atlantic Canada. **Also known as:** fire cherry.

American Mountain-ash

Sorbus americana

Height: to 10 m
Leaves: 15–20 cm long, alternate, pinnately compound with 11–17 lance-shaped, toothed leaflets, each 4–10 cm long
Flowers: 6 mm wide, white, 5 round petals, in rounded clusters
Fruit: bright orange-red pomes, to 1 cm wide, in clusters

American mountain-ash is a native shrub, but it is interspersed with the similar European mountain-ash (*S. aucuparia*) that has spread throughout the region. Both species produce a bounty of red fruit in autumn that remains on the tree through winter, providing food for wildlife. **Where found:** moist soils along roadsides and in open woodlands, valleys, coniferous forests and transitional forests; from Newfoundland south. **Also known as:** rowan tree, dogberry.

Downy Serviceberry

Amelanchier arborea

Height: up to 12 m
Leaves: 4–10 cm long, alternate, oval, finely sharp-toothed
Flowers: 1.5–2.5 cm across, white, 5 long, narrow petals, in elongated, drooping clusters
Fruit: dark reddish purple, dry, berry-like pomes, 6–10 mm wide

Serviceberry's beautiful white flowers are among the first of spring. • The name "serviceberry" is derived from "sarvissberry," which in turn comes from the Old English "sarviss." "Downy" refers to the woolly young leaves. The alternate name "shadbush" refers to the shad fish migration, which occurs about the same time this shrub blooms. **Where found:** dry, often sandy woods, rocky sites and transitional forests; from Newfoundland to NS. **Also known as:** smooth serviceberry, smooth juneberry, shadbush.

Steeplebush

Spiraea tomentosa

Height: to 1 m
Leaves: 3–5 cm long, alternate, oblong to lance-shaped, tapered at both ends, sharply toothed, white-woolly beneath
Flowers: small, pink to red, 5 petals, in steeple-shaped clusters
Fruit: small, dry, brown capsules with dense, woolly hairs

This small, woody shrub often forms dense, inconspicuous colonies when not in flower, but when it blooms, it is one of the most striking plants in its habitat, bedecked in eye-catching, showy, colourful flower clusters. • Grouse eat the young leaves, and deer also browse on this shrub. **Where found:** wet meadows, open, sunny areas in damp soils and along roadsides; throughout the Maritimes. **Also known as:** hardhack.

Rhodora

Rhododendron canadense

Height: 0.5–1.2 m
Leaves: 2–6 cm long, narrow, oval, grey-green, hairy beneath
Flowers: 2–3 cm wide, pink to lavender (rarely yellow), 5 petals, the top 3 fused almost to the end, in clusters of 2–6 at branch ends, open before leaves emerge
Fruit: orange-brown capsules, 1–1.2 cm long

The abundant blooming of rhodora across lowland wetlands in early spring is a sure sign that summer is around the corner. The green foliage does not appear until after the blooms are spent, so the effect is an unbroken expanse of purple flowers. **Where found:** mostly acidic soils, often in damp boggy sites; wet, somewhat open places including thickets along streams, swamps and lakes, coastal marshes and the edges of moist woodlands; from NL south. **Also known as:** rose tree.

Shrubby Cinquefoil

Dasiphora floribunda

Height: to 1.5 m
Leaves: grey-green, pinnately compound, with 3–7 narrowly oblong leaflets, each 2 cm long, edges often rolled under
Flowers: 1.5–3.2 cm across, yellow, saucer-shaped, 5 petals, single or in small clusters at branch tips
Fruit: hairy achenes, 2 mm long, in compact clusters

This hardy, deciduous shrub is widely used as an ornamental in gardens and public places. It also provides erosion control, especially along highways. • In the wild, this shrub's presence often indicates high-quality habitat. **Where found:** wet prairies and fens and rocky shores; throughout the Maritimes. **Also known as:** *D. fruiticosa, D. fruticosa* ssp. *floribunda, Pentaphylloides floribunda, P. fruticosa, Potentilla floribunda, P. fruticosa.*

Black Raspberry

Rubus occidentalis

Height: under 2 m
Leaves: 5–13 cm long and wide, light green above, much paler below,
palmately compound, with 3–5 irregularly toothed leaflets
Flowers: 2.5 cm wide, white, 5 petals, in clusters of 3–7
Fruit: black, seedy raspberries, 1.25 cm across

This shrub provides cover for wildlife in winter, and the fruit is enjoyed by both humans and animals. The tender young shoots may be eaten raw once the prickly outer layer has been peeled off. Fresh or completely dried raspberry leaves make excellent tea, but wilted leaves can be toxic. • The numerous species of blackberries and raspberries in the genus *Rubus* can be difficult to separate. This distinctive species has strongly whitened, or glaucous, stems. **Where found:** thickets, clearings and open woodlands; throughout Atlantic Canada.

Black Chokeberry

Photinia melanocarpa

Height: under 2 m
Leaves: 2.5–7.5 cm long, elliptical, shiny and dark green above, finely toothed
Flowers: 1.25 cm across, white, 5 petals, in clusters at ends of twigs
Fruit: shiny, black pomes, 6 mm across

Black chokeberry has showy flower clusters and leaves that turn a beautiful orange-red in autumn. It is less conspicuous when not in flower or fruit, but look closely for the elongated, raised, reddish glands on the upper midribs of the leaves. • This plant grows in conditions that vary from acidic tamarack bogs to dry prairies, and it can tolerate pollution, salt and drought. **Where found:** varied habitats including acidic sites, bogs, lakeshores, dunes, thickets, dry prairies and old fields; throughout Atlantic Canada.

Swamp Rose

Rosa palustris

Height: 0.6–2 m
Leaves: alternate, pinnately compound, with 7 finely toothed,
elliptical leaflets, each 2.5–5 cm long
Flowers: up to 5 cm wide, pink, 5 petals, solitary or in small clusters
Fruit: red, smooth hips, 1.25 cm long

As its name implies, swamp rose commonly grows in swampy, wet areas and will even survive in standing water. It is distinguished from other roses by its large, pink flowers, compound leaves with 7 leaflets and downcurved thorns. • Rose hips are rich in vitamins A, B, C, E and K. Three hips can contain as much vitamin C as an orange. **Where found:** swamps and bottomlands; throughout QC and the Maritimes.

Bunchberry

Cornus canadensis

Height: 5–20 cm
Leaves: 2–6 cm long, opposite, evergreen, elliptical, pointed at tip, deeply veined, in whorls of 4–6
Flowers: tiny, greenish white to purplish, in dense clusters of 5–15, surrounded by 4 greenish white, showy, petal-like bracts
Fruit: round, red, berry-like drupes, 6 mm across

This small shrublet has miniature clusters of tiny, whitish blooms surrounded by showy bracts (modified leaves) that look like petals. The gleaming white bracts attract insects and provide good landing platforms for pollinators to get at the flower nectar. • The edible drupes are not very flavourful, but the crunchy, poppy-like seeds are enjoyable. **Where found:** dry to moist sites and transitional forests; from Labrador south.

Red-osier Dogwood

Cornus sericea

Height: 0.5–3 m
Leaves: 2–8 cm long, opposite, oval to lance-shaped, pointed at tip, prominently veined
Flowers: less than 6 mm wide, white to greenish, in dense, flat-topped clusters at branch tips
Fruit: white, berry-like drupes, 5–6 mm across

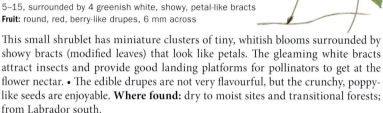

This attractive, hardy, deciduous shrub has distinctive purple to red branches with white flowers in spring, red leaves in autumn and white, berry-like fruits in winter. • A similar species, alternate-leaved dogwood (*C. alternifolia*) has green bark and is sometimes called green-osier dogwood. • An "osier" is a flexible branch that may be used to weave baskets and such. **Where found:** moist sites, transitional forests; from NL south. **Also known as:** *C. stolonifera.*

Winterberry

Ilex verticillata

Height: 0.9–3 m
Leaves: 3.5–9 cm long, alternate, elliptical, leathery
Flowers: 6–13 mm across, yellowish to greenish white, 4–8 petals, solitary or in small clusters
Fruit: round, glossy, red, berry-like drupes, 6–8 mm across

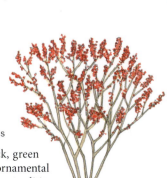

A member of the holly family, winterberry's thick, green foliage and bright red "berries" make it a popular ornamental in winter, particularly at Christmas. • The leaves were tradition-ally browned and then used as a tea substitute. • The berries are toxic and can cause nausea, diarrhea and vomiting. **Where found:** wet areas, swamps, damp woods, pond edges and transitional forests; east to NS and south to New England. **Also known as:** Canada holly, coralberry.

Staghorn Sumac

Rhus typhina

Height: to 9 m
Leaves: 30–50 cm long, alternate, pinnately compound with 11–31 dark green, lance-shaped leaflets; central stalks reddish, hairy
Flowers: 3–5 mm across, yellowish green, in dense, erect, cone-shaped clusters, 15–20 cm long
Fruit: red, hairy drupes, 3–5 mm across, in dense, erect, cone-shaped clusters

Showy red fruit clusters and colourful autumn leaves make staghorn sumac a favourite ornamental. In winter, the wide, woolly branches resemble velvet-covered deer antlers, inspiring the name "staghorn." • Delicious pink "lemonade" can be made by soaking the crushed fruit in cold water to remove the hairs, then adding sugar. **Where found:** open, often disturbed sites, typically on dry, rocky or sandy soil; roadsides, abandoned fields and transitional forests; east to NS and south to New England.

Poison Ivy

Toxicodendron radicans

Height: 10–25 cm, trailing or erect
Leaves: 18–25 cm long, alternate, compound with 3 shiny, oval, irregularly toothed leaflets, each 5 cm long
Flowers: 3 mm across, greenish white, 5 petals, in loose clusters
Fruit: round, greenish white berries, 6 mm across, in hanging clusters

This species is the one plant that anyone venturing outdoors should learn to recognize. Identification difficulties are compounded by its variable growth habit; it can appear as trailing groundcover, a small, erect shrub or a vine climbing high into trees or on other objects. A brush with poison ivy can cause a severe allergic reaction, obvious in an itchy rash and swelling. To hyper-responders, contact can even be life threatening. **Where found:** various dry to moist upland sites; throughout Atlantic Canada. **Also known as:** *Rhus radicans*.

Buttonbush

Cephalanthus occidentalis

Height: up to 6 m
Leaves: 6–15 cm long, opposite or rarely in whorls of 3, elliptical, glossy and dark green above, paler beneath, smooth edges
Flowers: small, creamy white, fragrant, funnel-shaped with 4 petal lobes, protruding style, in dense, spherical heads, 2–3.5 cm across
Fruit: reddish green to red-brown nutlets, in a dense cluster, 1–2 cm across

This species often forms a distinctive wetland plant community known as a buttonbush swamp. The shrubs can grow in fairly deep water, sometimes choking woodland pools. • Traditionally, the bark was used for stomach ailments and toothaches, as an eyewash for inflamed or irritated eyes and as a quinine substitute to treat malaria. **Where found:** wooded wetlands, marsh borders, edges of ponds and lakes, and transitional forests; from QC to NS and south.

Nannyberry

Viburnum lentago

Height: 4–7 m
Leaves: 6–10 cm long, opposite, elliptical with a sharply pointed tip, finely sawtoothed, smell unpleasant when bruised
Flowers: 4–8 mm across, white, 5-lobed corolla, in round-topped clusters 7.5–13 cm wide
Fruit: red to bluish black, berry-like drupes, 8–12 mm long, in open clusters

This shrub's summer leaves are shiny and green above, and yellow-green with tiny, black dots beneath. Its true beauty is revealed in autumn, when the leaves turn shades of purple, red or orange and the fruits ripen to deep red, blue or almost black. • Although the fragrance of the fruits has been compared to dirty socks, their flavour is sweet and tasty. **Where found:** shady and moist sites, mixed woodlands, open pastures, streambanks and along lakes and roadsides; east to NB.

Common Elderberry

Sambucus nigra ssp. *canadensis*

Height: up to 5 m, normally much shorter
Leaves: 10–30 cm long, opposite, pinnately compound, with 5–11 (usually 7) slender-pointed, sharply toothed leaflets
Flowers: tiny, white, star-shaped, fragrant, in flat-topped clusters
Fruit: purplish black, berry-like drupes, 6 mm across, on rose red stalks, in flat-topped clusters

Common elderberry's large, showy flower clusters make it conspicuous in early summer along county roadsides. • The fruits can be made into jam, jelly, pies and wine, but the raw berries are unpalatable and toxic (cooking destroys the toxins). • The leaves, bark and roots of this plant contain cyanide and are poisonous. **Where found:** moist, sunny sites, abandoned fields, woodland clearings and streambanks; throughout the Maritimes. **Also known as:** *S. canadensis.*

Bayberry

Morella pensylvanica

Height: 1–5 m
Leaves: 2–7 cm long, alternate, oval to oblong, leathery, aromatic when crushed
Flowers: tiny, yellowish green, in catkins 3–18 mm long
Fruit: round, blue to grey berries, 3–5 mm across, covered with whitish wax, in tight clusters

The leaves and twigs of this aromatic shrub are collected for their fragrance, and the leathery leaves can be used in cooking and give off a spicy aroma. • This is a beneficial shrub for the local ecology. It provides forage for wildlife but also adds nitrogen-fixing bacteria to the soil. **Where found:** dry sites, sandy soils, protected dunes, open and rocky woodlands and abandoned fields; from the Maritimes south. **Also known as:** candleberry; *Myrica pensylvanica.*

Wild Cucumber

Echinocystis lobata

Height: to 8 m, climbing vine
Leaves: 8 cm long, palmately lobed with 3–7 sharply triangular lobes
Flowers: 1.3–1.6 cm across, white, star-like with 6 long, narrow petals, in long, erect clusters
Fruit: 2.5 cm long, green, oval, fleshy capsules with numerous, soft prickles

The wild cucumber may have an appetizing-sounding name, but the fleshy fruits are inedible. • The bizarre, densely prickly fruits are buoyant, an excellent adaptation for water dispersal during floods. • This odd but showy vine can be quite common, especially in moist soils of alluvial thickets. **Where found:** wet to moist sites in thickets and along streambanks and roadsides; throughout the Maritimes.

Bittersweet

Celastrus scandens

Height: to 4.5 m or longer, climbing vine
Leaves: 5–10 cm long, oval to lance-shaped, pointed tip, finely serrated
Flowers: 4 mm across, green, 5 petals, in terminal clusters
Fruit: yellowish orange capsules, 8 mm across, open to expose scarlet, berry-like arils

Bittersweet often blends with other tangles of vegetation but becomes conspicuous when the bright orange and red fruits ripen. The showy "berries" provide winter food for birds, rabbits and squirrels but are poisonous to humans. All parts of the plant, including the berries, are potentially toxic and should not be consumed, especially by pregnant or nursing women. **Where found:** thickets, fencerows, woodland edges and riverbanks; throughout the Maritimes.

Riverbank Grape

Vitis riparia

Height: to 25 m or longer, climbing vine
Leaves: 7–20 cm long, alternate, 3-lobed (some unlobed), coarsely toothed
Flowers: tiny, greenish, fragrant, in compact pyramidal clusters
Fruit: spherical, waxy, black berries (grapes), 10–12 mm across, in hanging clusters

The small, tart fruits of this plant are juicy and flavourful, especially after the first frost. • Riverbank grape is a key parent species in breeding modern grape varieties, such as the Concord grape, that are disease and cold resistant. • Do not confuse this plant with the similar-looking Canadian moonseed (*Menispermum canadense*), which is poisonous; moonseed leaves have smooth rather than toothed edges, and the fruit has only a single seed. **Where found:** moist thickets and woodlands; throughout the Maritimes.

HERBS, GRASSES, FERNS & SEAWEEDS

Herbs include all non-woody, flowering plants that are not grass-like (grasses, sedges and rushes are all graminoids). They are often perennials that grow from a persistent rootstock, but many are short-lived annuals. A great variety of plants are herbs, including all our spring wildflowers, several flowering wetland plants, herbs grown for food or medicine and numerous weeds. Many herbs are used for adding flavour to foods and in herbal remedies, aromatherapy and dyes. Culinary herbs are typically made from the leaves of non-woody plants, but medicinal herbs are made from the flowers, fruit, seeds, bark or roots of both non-woody and woody plants. Herbs are also vital to the ecology of the plant communities in which they occur as food sources for pollinating insects and other animals, host plants for moths and butterflies, nest material for birds and cover for many animal species.

The herbs illustrated here are only the most frequent and likely to be seen. Of the many herbaceous plants in Atlantic Canada, just a sampling from the various plant families appears in this guide. We have included common, widespread, ecologically important species, as well as representatives found in the various habitats of forests, coastlines, open fields, low valleys and high elevations. These species should provide a good starting point for those wishing to delve further into the spectacular and diverse flora of our region.

Ferns are flowerless plants with feathery or leafy fronds and reproduce by means of spores. These plants first appeared in the fossil record 360 million years ago, though many of the current species did not evolve until the late Cretaceous period, about 145 million years ago. Some fern species can be gathered as food (fiddleheads) or medicine, or used as ornamental plants and for remediating contaminated soils, whereas others are considered weeds.

Seaweeds are algae and can be classified into 3 major groups: green, red and brown. They absorb all the required fluids, nutrients and gases directly from the water and, unlike terrestrial plants, do not require an inner system for conducting fluids and nutrients. However, seaweeds do contain chlorophyll to absorb the sunlight needed for photosynthesis. They also contain other light-absorbing pigments, which give some their red or brown colouration. Instead of roots, seaweeds have "holdfasts" to anchor them to the sea floor, and many seaweeds have hollow, gas-filled floats that help to keep the photosynthetic structures of these organisms buoyant and close to the water's surface so that they can absorb sunlight. Seaweeds provide food and shelter for marine animals, and dense, underwater seaweed "forests" are an important part of many marine ecosystems.

Lilies & Irises
pp. 172–174

Orchids
pp. 175–176

Ginger
p. 177

Buttercups
pp. 177–178

Bloodroot
p. 179

Dutchman's Breeches
p. 179

Pokeweed
p. 179

Smartweed
p. 180

Pitcher-plant
p. 180

Shinleaf
p. 180

Wood Sorrel
p. 181

Sundew
p. 181

Violet
p. 181

Sea Rocket
p. 182

Indian-pipe
p. 182

Primroses
pp. 182–183

Saxifrages
p. 183

Roses
p. 184

Twinflower
p. 184

Legumes
p. 185

Water-milfoil
p. 185

Evening-primroses
pp. 186–187

Beach Heath
p. 187

Touch-me-not
p. 187

Ginsengs
p. 188

Parsleys
pp. 188–191

Indian-hemp
& Milkweed
p. 191

Nightshade
p. 192

Phloxes
p. 192

Oysterleaf
p. 192

Vervain
p. 193

Mints
p. 193

Figworts
p. 194

Bladderwort
p. 194

Bellflowers
p. 195

Madders
pp. 195–196

Asters
& Sunflowers
pp. 196–200

Arrowhead
p. 200

Arums
pp. 200–201

Duckweed
p. 201

Pickerelweed
p. 201

Cattails, Grasses
& Sedges
pp. 202–204

Ferns
pp. 204–205

Sea Lavender
p. 205

Eel Grass
p. 205

Algaes
& Seaweeds
pp. 206–207

Canada Lily

Lilium canadense

Height: 1.8 m
Leaves: 8–15 cm long, lance-shaped, in whorls of 4–12
Flowers: 8 cm across, 6 tepals, yellow or orange to red, heavily spotted, nodding on long stalks
Fruit: erect, cylindrical capsules, 2–4 cm long

The petals and sepals of lilies are very similar and are collectively called "tepals." • The flowers of Canada lily vary in colour but are most often yellow. The tepals either spread or curl backward as the flowers bloom. • The flowers, seeds and bulbs of lilies have all been used traditionally as both food and medicine. **Where found:** wet meadows, moist, rich woodlands, marshes, swamps and along wet roadsides and railway tracks; from QC to NS. **Also known as:** wild yellow lily, meadow lily.

Painted Trillium

Trillium undulatum

Height: 20–50 cm
Leaves: 5–12.5 cm long, oval, tapering to a point, bluish green, waxy, prominently veined, in a whorl of 3
Flowers: 5–6.5 cm across, 3 petals, white with a dark pink, V-shaped stripe near the base of each petal, wavy margins, solitary
Fruit: shiny, red, ovoid berries, 1–2 cm long

Everything about this plant seems to be in threes—3 leaves, 3 petals, 3 sepals—which is what the name "trillium" refers to. • The white petals are "painted" with bold, pink brushstrokes, and the scientific name *undulatum* refers to the wavy or undulate margins of the petals. **Where found:** cool coniferous forests and acidic soils; east to NS south.

Yellow Trout-lily

Erythronium americanum

Height: 10–25 cm
Leaves: 10–20 cm long, 2 near the base of the stem, opposite, elliptical, green-mottled purplish brown
Flowers: 2.5–5 cm across, 6 backward-curved tepals, yellow, often red-spotted inside, solitary, nodding on long stalks
Fruit: flat, rounded capsules, 12–15 mm long

This lily was likely named for its leaf markings, which resemble the patterns on the brook trout. • The edible underground bulbs were collected as a food staple by many Native groups. All lilies have a degree of toxicity, and some are deadly poisonous, so it took astute observation and skilled preparation to use lilies for food and medicine. **Where found:** shaded and moist meadows and transitional forests; throughout the Maritimes. **Also known as:** dogtooth violet.

False Solomon's-seal

Maianthemum racemosum ssp. racemosum

Height: 30–90 cm
Leaves: 7.5–15 cm long, alternate, in 2 rows, elliptical, finely hairy beneath, prominent parallel veins
Flowers: 3 mm across, white, numerous in a dense terminal cluster
Fruit: round, red berries (may be green with red splotches), 6 mm across

In traditional medicine and alchemy, Solomon's-seal was thought to have mystical as well as healing properties, and it is still used today by herbalists. False Solomon's-seal was used for its similar appearance, but it is a different plant from the former, which is a *Polygonatum* species. **Where found:** moist, typically deciduous woods and clearings; from QC to NS. **Also known as:** Solomon's plume; *Smilacina racemosa*.

Canada Mayflower

Maianthemum canadense

Height: 5–25 cm
Leaves: 2–8 cm long, alternate, broadly heart-shaped to oval, pointed at tip, somewhat hairy
Flowers: 4 mm across, white, 4-parted, in clusters at stem tips
Fruit: hard, brown-spoltched, green berries, 3 mm across, becoming red and soft with age

Spreading by underground rhizomes, this flower often forms dense carpets of green leaves adorned with delicate flowers in spring, hence the name. • Many parts of this herb were used traditionally for food and medicine. The fruits are edible but should only be eaten in small amounts. **Where found:** upland woods, clearings and transitional forests; throughout Atlantic Canada. **Also known as:** wild lily-of-the-valley, false lily-of-the-valley, heartleaf, tobacco berry, bead ruby.

Bluebead Lily

Clintonia borealis

Height: 15–38 cm
Leaves: 12.5–20 cm long, basal, 2–5 (usually 3), oblong to elliptical, glossy, dark green
Flowers: 2–2.5 cm long, greenish yellow, bell-shaped, nodding, in clusters of 3–8
Fruit: ellipsoid, bright blue berries, 13 mm long

The shiny, almost metallic-looking, blue berries inspired the name for this lily. They are attractive but poisonous. • The genus name honours DeWitt Clinton (1769–1828), a New York governor who was a respected naturalist but also responsible for the construction of the Erie Canal. **Where found:** shady, moist woodlands, acidic soils and transitional forests; throughout Atlantic Canada. **Also known as:** balsam bell, bear plum, calf corn, Canada mayflower, cow tongue, wood lily, yellow clintonia.

Indian Cucumber Root

Medeola virginiana

Height: 30–75 cm
Leaves: 2.5–7.5 cm long, in 1–2 whorls of 3–11 leaves, narrowly elliptical, tapered at both ends, veined
Flowers: 13 mm long, greenish yellow, 6 backward-curved tepals, nodding on long stalks, in clusters of 3–9
Fruit: round, dark bluish purple berries, 8 mm across

The crisp, juicy rhizome of this plant tastes and smells quite like a cucumber. It is white and brittle, with much the same texture as a cucumber, as well. It can even be pickled! • The berry-like fruits are cherished by birds. • Flowers only appear on plants with 2 tiers of leaves and bloom from May to June. **Where found:** moist woodlands and hardwood and mixed coniferous-hardwood forests; from QC to NS south to Florida.

Sessile Bellwort

Uvularia sessilifolia

Height: 15–30 cm
Leaves: 4.5–7.5 cm long, alternate, sessile, narrowly elliptical
Flowers: 2.5 cm long, pale yellow, bell-shaped, 6 tepals, solitary, nodding
Fruit: sharply 3-angled capsules similar to a beechnut, 1.5–2 cm long

This common woodland flower is named for the shape of its flowers, which resemble the uvula in the throat. Because of this, early homeopaths surmised that the plant might be used to cure sore throats, which was incorrect. • The flowers have a rich, spicy fragrance, but this does not translate into flavour. **Where found:** deciduous woods, thickets and transitional forests; throughout the Maritimes. **Also known as:** wild oats, cornflower, straw lily, sessileleaf bellflower.

Northern Blue Flag

Iris versicolor

Height: 20–90 cm
Leaves: 10–80 cm long, 1–2.5 cm wide, basal, lance-shaped, folded along the midrib
Flowers: 6–7 cm wide, pale to dark violet-blue, 3 spreading petals with a yellowish spot near the base, 3 erect sepals, in clusters of 2–5
Fruit: beaked, oblong capsules, 3–5 cm long

Blue flags are the wild stock of domestic irises. The pale to deep blue flowers have the classic 3 backward-curving, purple-lined sepals and 3 erect, narrower inner petals (sepals). • Northern blue flag is the provincial flower of Québec, resembling the fleur-de-lys. • All parts of this plant are toxic and contain an acrid, resinous substance called irisin—from which the name iris derives. **Where found:** wet, open sites, roadsides and marshes; throughout Atlantic Canada.

Arethusa

Arethusa bulbosa

Height: 10–40 cm
Leaves: 4–20 cm long, single, grass-like
Flowers: 2–4 cm long, pale pink to magenta, erect sepals and petals, whitish pink lip has wavy margins, a yellow, hairy center and dark pink spots, solitary
Fruit: erect, ovoid capsules, 15–25 mm long

Arethusa was a beautiful Greek water nymph, and this flower does her no disservice, exemplifying delicate beauty. • Arethusa is indeed a temptress, luring inexperienced bees to her lovely colours with little reward of nectar; bees soon learn to avoid her, but it is early spring pollination, when bees are naïve, that counts. **Where found:** acidic soils, bogs, peat or sphagnum moss and lakesides; throughout the Maritimes. **Also known as:** swamp pink, dragon's mouth.

Grass-pink

Calopogon tuberosus

Height: to 65 cm
Leaves: 15–60 cm long, basal, 1–2, lance-shaped, strongly ribbed
Flowers: 2–3 cm wide, pale pink to bright magenta, uppermost petal narrow with a triangular tip and a patch of orange-yellow bristles, 2 lateral petals and 3 sepals widely spreading
Fruit: erect, cylindrical to ellipsoid capsules, 1.5–3 cm long

The genus name *Calopogon* means "beautiful beard" in Greek, which on this orchid refers to the hair-like projections on the lip of the flower. The flower's intricate architecture is designed to aid in pollination—when a bee lands on the flower, pollen is brushed all over its back, or pollen from a previously visited orchid is brushed off. **Where found:** wet meadows, fens and open seepage slopes; throughout the Maritimes. **Also known as:** bearded pink, meadow gift.

Calypso

Calypso bulbosa

Height: 7.5–20 cm
Leaves: 2–4 cm long, solitary, basal, ovate, dark green
Flowers: 1.5–2 cm long, solitary, rose purple, 3 erect sepals and 2 erect, twisted petals, 1 large slipper-like lower petal, yellow to whitish, streaked and spotted with purple and with a cluster of golden hairs
Fruit: erect, ellipsoid capsules, 2–3 cm long

This little orchid is special in our region, being the only member of its genus to brave Canada's northern climes. • The solitary basal leaf of this perennial withers after the flower blooms and is replaced with an over-wintering leaf. **Where found:** cool, damp, mossy woodlands, typically coniferous forests; across Canada to PEI. **Also known as:** fairy slipper.

Stemless Lady's-slipper

Cypripedium acaule

Height: 15–37.5 cm
Leaves: 10–20 cm long, 5–8 cm wide, basal, 2, opposite, sparsely hairy
Flowers: 3–6 cm long, yellowish brown to maroon sepals and petals and a large white, pink or magenta pouch, drooping, solitary at stem tips
Fruit: erect capsules, 4.5 cm long

The distinctive lower petal of this flower looks like a tiny slipper, giving this plant its demure name. It is not a slipper for the Cinderellas of the forest, however, as it is one of our largest native orchids and the provincial flower of Prince Edward Island. **Where found:** dry forests, pine forests, humus-rich and sandy soils and occasionally moist woodlands; from QC to Newfoundland and south. **Also known as:** pink lady's-slipper, pink moccasin-flower.

Large Purple Fringed Orchid

Platanthera grandiflora

Height: 60–120 cm
Leaves: to 20 cm long, 2–5, alternate, lance-shaped, keeled, reduced to bracts above
Flowers: 2.5 cm long, fragrant, showy, lavender to rose purple, 3 erect, oval sepals and 2 small, upward curved lateral petals, large lip petal has 3 deeply fringed lobes, flowers in a dense, cylindrical raceme
Fruit: ellipsoid capsules, 1.5–2 cm long

This lovely, deeply fringed orchid is delicately coloured and scented, but it is a hardy perennial despite its seeming fragility. • The sticky nectar from the fragrant flowers attracts moths, which inadvertently pollinate the flowers during their evening dining forays. **Where found:** cool, moist woods, wet meadows, swamps and transitional forests; from Newfoundland south.

Early Coral Root

Corallorrhiza trifida

Height: 8–35 cm
Leaves: reduced to bladeless bracts sheathing the stem
Flowers: 6–10 mm wide, yellowish white to greenish, 3–12 in a terminal cluster
Fruit: drooping capsules, 1 cm long, with many seeds

This native orchid is yellow to green in colour from stem to flower, with a somewhat translucent quality to the stem. The inner lip of the flower is a slightly contrasting white, but the flower does not need to be colourful—early coral root is saprophytic; it gets its nutrients from the leaf litter on the forest floor and does not require the assistance of pollinators. **Where found:** bogs, upland woods and deciduous and coniferous forests; throughout Atlantic Canada.

Wild Ginger

Asarum canadense

Height: 15–30 cm
Leaves: 7.5–15 cm wide, opposite, heart-shaped, downy
Flowers: 3–4 cm across, tan to purple, urn-shaped with 3 long, pointed sepals, solitary, often lying on the ground
Fruit: inconspicuous capsules

Wild ginger flowers lie on the ground, hidden under the large leaves, and are pollinated by crawling as well as flying insects.
• This plant looks and smells like commercial ginger and can be used in much the same way. The roots can be used to flavor foods and drinks and to treat certain ailments. The leaves have antifungal and antibacterial properties. • Wild ginger and commercial ginger (*Zingiber officinale*) are not even closely related. **Where found:** rich woodlands, usually deciduous, and transitional forests; from QC to NB south to South Carolina.

Marsh-marigold

Caltha palustris

Height: 20–60 cm
Leaves: 2–13 cm long, round to kidney-shaped, heart-shaped at base, serrated, waxy
Flowers: 2–4 cm across, bright yellow, 5–9 petal-like sepals, solitary
Fruit: clusters of 6–12 curved, beaked pods (follicles), 15 mm long

This sunny yellow flower brightens up its swampy habitat. Although the genus name *Caltha* is Latin for "marigold," this plant is a member of the buttercup family. • Marsh-marigold has traditionally been used for a variety of medicinal and culinary purposes, but do not eat this plant unless you know precisely how to prepare it. It contains poisonous glycosides and protoenemonin and helleborin poisons. Contact with the skin may cause blistering. **Where found:** marshes and other wet areas, and boreal forest; throughout Atlantic Canada.

Hepatica

Anemone americana

Height: 10–15 cm
Leaves: 2.5–7.5 cm across, basal, dark green, leathery, 3-lobed with rounded tips
Flowers: 1.5–2.5 cm across, pale pink to lavender or white, 5–12 petal-like sepals, numerous yellow stamens, subtended by 3 green bracts
Fruit: hairy, seed-like achenes

The 3-lobed leaves characteristic of this plant were thought to resemble the shape of the liver, which is where the plant's name derives from.
• Bees and flies are the main pollinators, but ants disperse the seeds later in the season. **Where found:** rich, upland woods, acidic soils and mixed forests; from QC to NS and south. **Also known as:** *Hepatica americana*.

177

Bristly Buttercup

Ranunculus hispidus

Height: 30 cm
Leaves: 5–10 cm long, basal and alternate on the stem, on long, hairy stalks, deeply divided into 3 toothed or lobed leaflets
Flowers: 2–2.5 cm across, 5 shiny, yellow petals, 5 yellow-green sepals, many yellow stamens, solitary on hairy stalks
Fruit: smooth, flattened achenes, 3.5 mm long

The genus name *Ranunculus* comes from the Latin, *rana*, meaning "frog," because buttercups tend to grow in moist places; *hispidus* means "covered in stiff, coarse hairs" and refers to the hairs that cover the entire plant. • The seeds are eaten by birds and mammals such as grouse and chipmunks, but the foliage is toxic to mammals. **Where found:** medium-moist to moist woodlands, slopes and valleys, as well as swamps and wetlands; throughout Atlantic Canada. **Also known as:** hispid buttercup.

Goldthread

Coptis trifolia

Height: 7.5–15 cm
Leaves: 2.5–5 cm long, basal, evergreen, shiny, divided into 3 wedge-shaped, toothed leaflets
Flowers: 1–1.5 cm across, white with a yellow centre, 5–7 petals, solitary
Fruit: 3–9 dry, long-stalked, beaked pods, splitting open along one side

One wonders why, with its dark evergreen leaves and solitary white flowers, this plant is called goldthread. The yellow, thread-like underground stem is the hidden treasure of this species; it has many medicinal properties and is used to treat mouth sores (lending it the common name "cankerroot") and to make an eyewash to sooth eye irritations. **Where found:** cool woodlands, swamps, bogs and transitional forests; throughout Atlantic Canada. **Also known as:** cankerroot; *C. groenlandica.*

Red Baneberry

Actaea rubra

Height: 20–80 cm
Leaves: 2–10 cm long, alternate, 2– times divided into 3s, coarsely toothed
Flowers: 6 mm across, white, 5–10 petals, 3–5 sepals, numerous stamens, in rounded clusters
Fruit: clusters of glossy, red (occasionally white) berries on slender stalks

Baneberries are toxic but poisoning is rare because of the bitterness of the berries. The high toxicity of the plant is attributed to either the glycoside ranunculin or to an as-yet-unidentified essential oil. • Herbalists have taken advantage of medicinal properties in the roots of this plant, prescribing it as an antispasmodic, anti-inflammatory, vasodilator and sedative. **Where found:** woods and thickets, streambanks, swamps and deciduous, mixed coniferous and transitional forests; from Newfoundland south.

Bloodroot

Sanguinaria canadensis

Height: 25–40 cm
Leaves: 10–20 cm long, 1, round, pale green, 3–9-lobed
Flowers: 2.5–5 cm across, 8–16 white petals, numerous
yellow stamens, solitary
Fruit: 2-parted capsule, pointed at both ends

Bloodroot "bleeds" a brilliant crimson sap when the
rhizome is cut. It is well known for its antimicrobial,
anti-inflammatory and antioxidant properties, and it was traditionally used to
treat any blood-related disease or malady. Bloodroot was also, more logically,
used as a red dye, and as a love charm or aphrodisiac. **Where found:** streambanks,
rich woodlands, thickets and transitional forests; throughout Atlantic Canada.

Dutchman's Breeches

Dicentra cucullaria

Height: 10–30 cm
Leaves: 7.5–15 cm long, trifoliate with finely divided leaflets
Flowers: 2 cm, white, pantaloon-shaped, hanging on an arching stalk
Fruit: oblong capsules, opening into 2 parts when mature

This plant's common name refers to the flowers, which resemble
the baggy breeches and yellow-orange sash worn by "Dutchmen."
• *Dicentra* species contain several isoquioline alkaloids that negatively affect the
nervous system if ingested. If cattle graze upon these plants, one of the most
common symptoms of poisoning is a staggering gait, which gives this plant the
alternate common name of staggerweed. **Where found:** fresh to moist hardwood
forests, rich woodlands and transitional forests; from QC through the Maritimes;
rare to absent in Newfoundland. **Also known as:** staggerweed, white hearts.

Pokeweed

Phytolacca americana

Height: to 3 m
Leaves: 30 cm, alternate, lance-shaped, smooth
Flowers: 6 mm across, greenish white, 5 petal-like sepals, numerous
stamens, in upright or drooping racemes
Fruit: dark purple, nearly round berries, 1 cm across, on pink stalks

Despite the berries, seeds, roots and mature stems and leaves
of this plant being very poisonous, it was somehow discovered that
the young shoots and leaves are edible after they have been boiled
multiple times. A less risky use of this plant was as a dye made from the red juice
of the berries. • **Where found:** damp, rich soils in clearings, pastures, thickets and
disturbed sites and along roadsides; from QC through the Maritimes. **Also
known as:** inkberry, American nightshade, pigeon berry, redweed.

Water Smartweed

Persicaria amphibia

Height: 0.5–1 m; to 3 m when aquatic
Leaves: 5–15 cm long, alternate, lance-shaped to elliptical, somewhat leathery
Flowers: 4–5 mm across, pink, 5-lobed, in dense, terminal clusters
Fruit: shiny, brown, lens-shaped achenes, 3 mm long

On the water, this plant has long (to 3 m) floating stems and forms large colonies that extend across the surface. Water smartweed can also be terrestrial, with hairier, lance-shaped leaves; the aquatic form has elliptical leaves that are more rounded to the tip. • The seeds are edible and are typically pounded into meal or flour. • This is an introduced species from Europe. **Where found:** low, wet habitats including ponds, swamps and damp meadows; throughout the Maritimes. **Also known as:** scarlet smartweed; *Polygonum amphibium*.

Pitcher-plant

Sarracenia purpurea

Height: 30–69 cm
Leaves: 10–30 cm long, basal rosette of curved, ascending, hollow "pitchers"
Flowers: 5 cm wide, deep purple (occasionally yellow), round, 5 curved petals, 5 spreading sepals, solitary on long stalks
Fruit: capsules with numerous laterally winged seeds

Pitcher-plant was declared the floral emblem of Newfoundland in 1954. It is a symbol of hardiness, resilience, adaptability and natural beauty. • This carnivorous plant lures its prey into the "pitchers," which have striking red veins and pungent nectar. It can digest almost any protein source that falls victim, including insects, spiders and even small amphibians. **Where found:** sphagnum bogs and wet meadows; from QC to Newfoundland south.

Shinleaf

Pyrola elliptica

Height: 8–30 cm
Leaves: 3–7 cm long, evergreen, basal, oval, rounded at tip
Flowers: 8–16 mm, waxy, white to pinkish, 5 oval, often green-veined petals, orange-tipped stamens, pale green style curves down and out like an elephant's trunk, in racemes of 3–16
Fruit: flattened, spherical capsules, 4–12 mm across

The pear-shaped leaves are one of the distinguishing characteristics of this plant, and the genus name *Pyrola* comes from the Latin *pyrus*, which means "pear-shaped." • The leaves and roots were used medicinally to make infusions, gargles and, particularly, poultices and plasters, which is the likely etymology of the common name. **Where found:** bogs, fens, swamps and moist to wet woodlands; throughout Atlantic Canada. **Also known as:** white wintergreen, waxflower.

Common Wood Sorrel

Oxalis montana

Height: 10–15 cm
Leaves: 5–6.3 cm, basal, long-stalked, palmately divided into 3 heart-shaped leaflets
Flowers: 1.3–2.5 cm across, white veined with pink, 5 petals, usually solitary
Fruit: tiny, hairy, seed-like capsules

Although common throughout our woodlands, this plant is difficult to grow in gardens—many have tried to include the pink to white flowers and clover-like leaves in their flowerbeds with limited success. • *Sorrel* is German for "sour" and refers to the tangy taste of the leaves, which are delicious in salads. The leaves are also rich in vitamin C and were once used to treat scurvy. **Where found:** damp, coniferous forests; throughout Atlantic Canada. **Also known as:** sour grass, sour clover.

Narrow-leaved Sundew

Drosera intermedia

Height: 10 cm
Leaves: to 6.5 cm long, mostly in a basal rosette, pale green to reddish, spoon-shaped, on long, narrow stalks, covered with reddish, sticky, gland-tipped hairs
Flowers: 6 mm wide, white, 5 oval petals, 1 to several on a stem to 10 cm tall
Fruit: small capsules with reddish brown seeds

This insectivorous plant produces sticky "dew" on the glandular hairs of the leaves to trap and digest its prey. Early experiments to find medicinal properties in this plant correctly assumed that this sticky substance might sooth coughs and tickly throats. Flavinoids in the dew have antispasmodic properties. Today, sundew is an ingredient in hundreds of registered medicines. **Where found:** wet, open sites, bogs, marshes and fens: throughout the Maritimes.

Blue Violet

Viola cucullata

Height: less than 20 cm
Leaves: 5–10 cm long, basal, heart-shaped, on long stalks
Flowers: 1.3 cm, violet-blue to white, usually darker at the centre, 5 petals, on tall stalks
Fruit: ovoid capsules with dark seeds

This lovely plant is the floral emblem of New Brunswick. More than just a pretty flower, it has a lengthy list of culinary, medicinal and practical uses. The flowers can be eaten raw or candied, and the leaves used as potherbs or thickeners. This plant is extremely rich in vitamins A and C. Obscure additional uses include being a substitute for litmus paper. **Where found:** moist meadows; throughout Atlantic Canada.

Sea Rocket

Cakile edentula

Height: 15–50 cm
Leaves: 2.5–5 cm long, 6–12 mm wide, fleshy,
oblong, lobed, edges wavy or toothed
Flowers: 6 mm across, pale lavender to white, 4 petals, in small clusters
Fruit: elongated seedpods (silicles), to 2.5 cm long

The rocket-shaped seed capsules for which this plant is named form after the pale lavender flowers finish blooming. The stems and leaves are fleshy and thick to store water and to withstand the desiccating effect of sand and sun. • Sea rocket is a pioneer species that colonizes sandy beaches. It helps to bind and stabilize the sand and add nutrients to the impoverished soil. **Where found:** above the high-tide line on sandy beaches; from southern NL south through the Maritimes.

Indian-pipe

Monotropa uniflora

Height: 5–30 cm
Leaves: to 1 cm long, alternate, scale-like
Flowers: 13–25 mm, white to salmon pink, 5 petals, narrowly bell-shaped, nodding
Fruit: ovoid capsules, 5–7 mm wide, becoming erect when mature

This unusual, fleshy, waxy perennial is easily mistaken for a fungus because its lack of chlorophyll imbues no green colouration. Rather than using photosynthesis, this parasitic species obtains nutrients from nearby plants through connections with mycorrhizal fungi. This adaptation also allows it to grow in darkness. **Where found:** woodland humus, dense, moist woodlands and transitional forests; throughout Atlantic Canada. **Also known as:** corpse plant, ghost flower, ice plant.

Fringed Loosestrife

Lysimachia ciliata

Height: 30–120 cm
Leaves: 5–12 cm long, opposite, oval, pointed at tip
Flowers: 1.2–2.5 cm across, bright yellow, deeply 5-lobed, bell-shaped,
1–3 on slender stalks
Fruit: egg-shaped capsules, 6 mm long

The name of this plant comes from the belief that it could calm, or loosen, strife. Legend says that the Sicilian king Lysimachus, successor to Alexander the Great, was once being chased by a mad bull; the king forced a sprig of loosestrife at the animal, and it was instantly calmed. Traditionally, loosestrife was used to soothe yoked animals and prevent them from fighting. **Where found:** moist or wet woodlands; throughout the Maritimes.

Starflower

Trientalis borealis

Height: 10–20 cm
Leaves: 4.5–10 cm long, lance-shaped, in whorls of 5–9 at stem tips
Flowers: 0.8–1.3 cm across, white, 5–9 pointed petals, solitary or 2–3 on slender stalks
Fruit: spherical, 5-valved capsules

This plant's pure white flowers against a background of dark leaves resemble stars in the sky. *Borealis* means "northern," and this species is a representative wildflower of our boreal forest. The leaves of this perennial herb wither back and disappear in autumn but reappear fresh and green in spring to provide the backdrop to a galaxy of flowers by May or June. **Where found:** cool, rich woodlands, peaty slopes, bogs and transitional forests; from NL south. **Also known as:** chickweek, evergreen, maystar.

Sea-milkwort

Glaux maritima

Height: 3–30 cm
Leaves: 5–25 mm long, opposite, fleshy, oval, blunt-tipped
Flowers: 4–5 mm long, white or pinkish, 5 petal-like sepals, solitary and stalkless at leaf bases
Fruit: spherical capsules, 2–3 mm long

Called milkwort for its use in infusions for nursing mothers, this plant is a member of the primrose family. • First Nations peoples cooked the rhizomes, and Europeans pickled the leaves. • The genus name *Glaux*, Greek for "bluish green," aptly describes the colour of the foliage, and *maritima* refers to this plant's coastal habitat. **Where found:** coastal tidelands; also inland salt marshes and wet meadows; throughout the Maritimes. **Also known as:** black saltwort.

Purple Saxifrage

Saxifraga oppositifolia

Height: 2–5 cm, cushion-like or trailing
Leaves: 3–6 mm long, scale-like, green-grey, succulent, in overlapping pairs
Flowers: 1 cm across, purple (rarely white), 5 petals, star-shaped, solitary
Fruit: brownish, 2-parted capsules

The Subarctic region of maritime Canada may be austere, but in spring it is awash with colour. The first flower to bloom, purple saxifrage is iconic, heralding the Arctic spring and coinciding with the calving of the northern caribou herds. • The edible flowers are considered a sweet treat by the Inuit. **Where found:** moist, calcium-rich, gravelly soils; Subarctic and alpine (to 1000 m) regions of NL, maritime QC including the Gaspé Peninsula, and may occur in NS. **Also known as:** French knot moss.

Common Strawberry

Fragaria virginiana

Height: up to 15 cm
Leaves: 2.5–10 cm long, divided into 3 coarsely toothed leaflets
Flowers: 1–2 cm across, white, 5 petals, in open clusters
Fruit: small, red strawberries dotted with achenes

Few things beat running into a patch of fresh wild strawberries. The fruit is delicious, and many animals enjoy it as well. • There are several species of strawberry throughout Canada, always distinguishable by the spreading runners, white flowers and, of course, sweet, red fruits. • This plant is the ancestor of 90 percent of our cultivated strawberries. **Where found:** dry fields and open forests; throughout Atlantic Canada.

Silverweed

Argentina anserina

Height: 15 cm
Leaves: 10–20 cm long, compound, divided into 7–21 coarsely toothed, hairy leaflets
Flowers: 1.5–2 cm across, bright yellow, 5–7 petals, solitary on long stems
Fruit: dry, flattened achenes, 1.5–2 mm long, in clusters

Silverweed's leaflets are covered in long, white, silky hairs, lending this plant the silvery appearance that gives this species its name. Until the 1990s, this plant was assigned to the genus *Potentilla*, a group of plants noted for their potency in numerous traditional medicines and topical treatments. **Where found:** gravelly shores, riverbanks, grassy meadows and roadsides; throughout Atlantic Canada. **Also known as:** silver cinquefoil; *Potentilla anserina*.

Twinflower

Linnaea borealis

Height: flowering branches 7.5–15 cm high, arising from semi-woody runners
Leaves: 1–2 cm long, opposite, evergreen, oval
Flowers: 6–15 mm long, whitish to pink, trumpet-like, fragrant, in pairs on Y-shaped stalks
Fruit: dry, egg-shaped nutlets, 2 mm long

The small, delicate pairs of pink bells are easily overlooked among other plants on the forest floor, but their strong, sweet perfume may draw you to them in the evening. • Hooked bristles on the tiny nutlets catch on fur, feathers or the clothing of passersby, who then carry these inconspicuous hitchhikers to new locations. **Where found:** moist, cool, open or densely shaded forests, shrub thickets, muskeg, bogs, rocky shorelines and transitional forests; throughout Atlantic Canada.

Lupine

Lupinus spp.

Height: to 75 cm
Leaves: to 7 cm long, grey-green, hairy, compound with multiple, pointed leaflets
Flowers: 1.5–2 cm long, showy, pea-like, yellow to purple, in long, loose clusters
Fruit: hairy pods, 2–6 cm long

These attractive perennials, with their showy flower clusters and fuzzy seedpods, enrich the soil with nitrogen. • The seedpods look like hairy garden peas, and children may incorrectly assume that they are edible. Many lupines contain poisonous alkaloids, and it is difficult to distinguish between poisonous and non-poisonous species. **Where found:** wet, open areas and disturbed sites; throughout Atlantic Canada.

Beach Pea

Lathyrus japonicus

Height: 30–150 cm, stems trailing
Leaves: 5–10 cm long, waxy, compound with 2–5 pairs of leaflets
Flowers: 3 cm long, pea-like, reddish purple to blue, in racemes of 2–7
Fruit: pods, 4–7 cm long

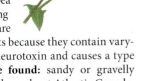

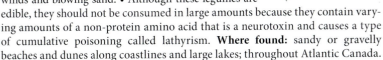

Long, curly tendrils cling to other plants as the beach pea spreads along the ground, anchoring itself against strong winds and blowing sand. • Although these legumes are edible, they should not be consumed in large amounts because they contain varying amounts of a non-protein amino acid that is a neurotoxin and causes a type of cumulative poisoning called lathyrism. **Where found:** sandy or gravelly beaches and dunes along coastlines and large lakes; throughout Atlantic Canada.

Eurasian Water-milfoil

Myriophyllum spicatum

Height: stems to 3 m long, aquatic
Leaves: 1.5–3.5 cm long, divided into numerous, feathery leaflets, in whorls of 3–5 on submerged stems
Flowers: 4–6 mm long, orange-red, clustered on a spike 5–15 cm long, rising above the water
Fruit: 4-lobed nutlets

As the name implies, this species was introduced from Eurasia; it arrived in North America between the 1950s and 1980s and has since become invasive. One of the adaptations that allows it to spread so successfully is its ability to grow from broken-off stems. It is often used in aquariums for its ease of growth, but it is difficult to keep in check. **Where found:** lakes, ponds, estuaries and other aquatic habitats; throughout Atlantic Canada.

Purple Loosestrife

Lythrum salicaria

Height: 1.5 m
Leaves: to 7.5 cm long, mostly opposite or whorled, dark green, lance-shaped
Flowers: 0.7–1 cm long, red-purple, usually 6 narrow, wrinkled petals, in spikes
Fruit: capsules with numerous seeds

Beautiful but noxious, purple loosestrife is an invasive weed on the most-wanted list for eradication in North America. It was introduced from Eurasia and took to its new land with enthusiasm. The biggest threats are its invasiveness in wetlands and its ability to hybridize with plants in the native *Lathyrum* genus. Purple loosestrife chokes waterways, pushing out native plants such as cattails. A single one of these plants can produce millions of seeds. **Where found:** wet areas; throughout Atlantic Canada. **Also known as:** kill weed, rosy strife.

Fireweed

Epilobium angustifolium

Height: 0.3–3 m
Leaves: 2–20 cm long, alternate, narrowly lance-shaped
Flowers: 2–5 cm across, pink to purple, 4 petals, in long clusters at stem tips
Fruit: linear pods, 5–7.5 cm long

Fireweed helps heal landscape scars such as burned forests by blanketing the ground with colonies of plants, producing a sea of deep pink flowers. The erect, linear pods split lengthwise to release hundreds of tiny seeds tipped with fluffy, white hairs. • Young fireweed shoots can be eaten like asparagus and the flowers added to salads. **Where found:** open, often disturbed sites and transitional forests; throughout Atlantic Canada.

Common Evening-primrose

Oenothera biennis

Height: 0.5–1.5 m
Leaves: 2–15 cm long, alternate, lance-shaped, slightly toothed
Flowers: 3–4 cm across, bright yellow, tube-shaped
Fruit: erect, cylindrical capsules, 2–4 cm long

The flowers of this well-named species open near dusk, bloom throughout the night and generally close by midmorning, making moths the prime pollinators. The flowers open amazingly quickly, going from shrivelled wisps to robust blossoms in just 15 to 20 minutes. • This plant is best known for evening-primrose oil, which is produced from the seeds. The oil has potent medicinal properties and is being tested as a treatment for a wide variety of conditions, mostly pain relief. **Where found:** dry, open sites, woods and meadows; throughout Atlantic Canada. **Also known as:** cure-all, fever-plant.

Enchanter's-nightshade

Circaea lutetiana

Height: 30–60 cm
Leaves: 5–10 cm long, opposite, oval, shallowly and irregularly toothed
Flowers: small, numerous, white or pinkish, 2 deeply notched petals, well spaced in terminal racemes
Fruit: small, nut-like pods

The name of Circe, the mythological Greek enchantress, was given to this charming, delicate wildflower perhaps for its own enchanting beauty or, according to some sources, because the sorceress made a love potion from this nightshade. Some claim that Circe used poisonous *Circaea* plants in her magic. • To truly appreciate these miniscule flowers, you'll have to use a magnifying lens. **Where found:** moist, rich woodlands; throughout Atlantic Canada.

Beach Heath

Hudsonia tomentosa

Height: 7.5–20 cm, mat forming
Leaves: 3 mm long, alternate, scale-like, grey-woolly, pressed close to the stem, overlapping like shingles
Flowers: 6 mm across, bright yellow, 5 petals, crowded near branch tips
Fruit: smooth, egg-shaped capsules

Beach heath is a pioneer species, able to take hold in poor and sandy soils, thus stablizing them. For this reason, it is often used in beach restoration. Tolerant of dry conditions but not salt spray, this plant is found farther back than the beach grasses in the foredunes. The small, hairy leaves and low growth height are adaptations to reduce water loss. • **Where found:** on sand dunes in the backdune areas of beaches and openings with poor soil; all along the coast. **Also known as:** false heather, sand golden heather, woolly hudsonia.

Spotted Touch-me-not

Impatiens capensis

Height: 50–150 cm
Leaves: 3–10 cm long, alternate, oval, serrated margins
Flowers: 2.5 cm long, orange-yellow, sac-like sepal is heavily spotted with reddish brown, flowers hang from thread-like stalks
Fruit: green capsules, 2 cm long, open explosively when touched

Exceptionally succulent, this plant will wilt in your hands like melting ice if picked, hence it begs you to "touch it not." • The seeds are enclosed in fleshy capsules and held by tightly coiled, elastic attachments. Press a ripe pod, and the seeds shoot forth explosively. Catch the seeds in your hand, pop them in your mouth and enjoy the taste of walnuts. **Where found:** moist, shaded woodlands, damp sites and lakesides; throughout Atlantic Canada. **Also known as:** jewelweed.

Dwarf Ginseng

Panax trifolius

Height: 10–20 cm
Leaves: 8 cm long, whorl of 3, palmately divided into
3–5 coarsely toothed leaflets
Flowers: tiny, whitish or yellow-green, 5 petals,
in a round cluster above the leaves
Fruit: yellow, berry-like drupes, with 2–3 seeds

Ginseng root has a rich 5000-year history of herbal use. According to traditional Chinese medicine, it regulates the balance between yin and yang, promotes health, vigour and long life, and is considered an aphrodisiac. In the West, ginseng root is known for its ability to boost the immune system, increase mental efficiency, improve physical performance and aid in adapting to stress. **Where found:** moist, rich woodlands; throughout the Maritimes.

Wild Sarsaparilla

Aralia nudicaulis

Height: 20–70 cm
Leaves: 30–60 cm long, single, compound, divided into 3, each division
with 3–5 oval to lance-shaped, finely toothed leaflets
Flowers: 2–3 mm across, greenish white, 5 petals, in clusters of 2–7
Fruit: purplish black berries, 6–8 mm across

The sweet, aromatic rhizomes of this plant were traditionally used to make tea, root beer and mead. The berries, generally inedible, have been also been used in making beer and wine. • The rhizomes were pulverized into poultices to heal wounds, burns and other skin ailments and to reduce swelling and inflammation. **Where found:** moist, shaded woods and transitional forests; throughout Atlantic Canada.

Snakeroot, Black Snake

Sanicula marilandica

Height: 40–120 cm
Leaves: 6–15 cm long, compound, divided into 5–7 double-toothed leaflets
Flowers: tiny, greenish white, 5-parted, in clusters of 12–25
Fruit: oval, seed-like schizocarps, 4–6 mm long

Snakeroot is reported to have sedative properties for soothing nerves and relieving pain. Mashed roots were once applied to snakebites. Native peoples considered this plant a powerful medicine and used it to treat many disorders. Caution must be taken, however, because the roots contain irritating resins and volatile oils. **Where found:** moist, deciduous woodlands; throughout Atlantic Canada.

Smooth Sweet-cicely

Osmorhiza longistylis

Height: 90 cm
Leaves: 1.5–9 cm long, alternate, compound, fern-like, serrated
Flowers: tiny, white, 5 petals, in branched clusters
Fruit: capsules with flattened nutlets

Two species of sweet-cicely are abundant in the region. The other is woolly sweet-cicely (*O. claytonii*), which is often hairier and lacks the strong anise scent of this species. • Smooth sweet-cicely was widely used medicinally by Native peoples. The leaves, stems and roots were consumed in infusions for stomach discomforts and kidney troubles. **Where found:** moist to dry woodlands; throughout Atlantic Canada.

Water Parsnip

Sium suave

Height: 50–100 cm
Leaves: alternate, pinnately divided into 5–17 slender, sawtoothed leaflets, 5–10 cm long
Flowers: tiny, white, in dense umbels 5–18 cm across
Fruit: seed-like schizocarps, 2–3 mm long, in pairs

All members of the carrot/parsnip family have umbels of tiny, clustered flowers. • The young stems and roots are edible, crisp and bitter like parsnips, and were eaten by Native peoples. • The flowerheads of this plant are poisonous. It is also very similar to the extremely poisonous water-hemlocks (*Cicuta* spp.). **Where found:** wet sites, fields and hillsides; throughout Atlantic Canada except Labrador.

Common Water-hemlock

Cicuta maculata

Height: 50–180 cm
Leaves: alternate, compound, pinnately divided 2–3 times into narrow, serrated leaflets
Flowers: tiny, white, 5 petals, clustered in small heads (umbellets), forming umbels to 15 cm wide
Fruit: flattened capsules (schizocarps) with ribbed nutlets

Among the most poisonous of our native plants, common water-hemlock contains deadly cicutoxin, which is concentrated in the rhizomes. Just a few small pieces may be fatal to a human; a mouthful can take down a horse. The poison acts upon the central nervous system and takes effect within 15 minutes of ingestion. **Where found:** wet soils, streambanks and rich woodlands; throughout Atlantic Canada. **Also known as:** spotted water-hemlock.

Scotch Lovage

Ligusticum scoticum

Height: 10–80 cm
Leaves: thick, firm, compound, divided into usually 9 egg-shaped, toothed leaflets, 2–6 cm long
Flowers: small, white to pinkish, clustered in small heads in compound umbels
Fruit: oblong, ribbed capsules, 7–8 mm long

Scotch lovage is used today for culinary purposes and herbal remedies. Considered a good source of vitamins A and C, it was traditionally eaten in salads, used to flavour soups, fish or meat, or the stalks were cooked like celery. • Washing lovage roots is superstitiously thought to bring on rainstorms. **Where found:** at elevations near sea level; upper beaches and coastal bluffs; throughout Atlantic Canada. **Also known as:** beach lovage.

Cow Parsnip

Heracleum sphondylium ssp. *montanum*

Height: 1–2.5 m
Leaves: 10–30 cm wide, alternate, compound, divided into 3 lobed, coarsely toothed leaflets
Flowers: tiny, white, in compound umbels 10–30 cm wide
Fruit: flattened, egg-shaped capsules (schizocarps), 7–12 mm long

The tiny flowers contrast with the overall largeness of this member of the carrot/parsley family. • Cow parsnip is edible and was a valuable staple to many First Nations groups. However, the plant can cause skin problems in photosensitive people because of irritative furanocoumarins. • Be careful not to confuse this plant with highly poisonous water-hemlocks (*Cicuta* spp.), which are similar in appearance. **Where found:** streambanks, moist slopes and clearings, upper beaches and marshes; throughout Atlantic Canada. **Also known as:** *H. lanatum, H. maximum.*

Marsh-pennywort

Hydrocotyle americana

Height: 5–13 cm
Leaves: 1–6 cm long, oval to round, scalloped edges
Flowers: 2 mm across, greenish white, 5 petals, in small bunches in leaf axils
Fruit: tiny, dry capsules, splitting into 2 seeds

This aquatic plant has leaves reminiscent of nasturtiums. The name "pennywort" comes from the European species, whose round leaves are the size of coins. • Historically, marsh-pennywort was sometimes used to treat skin ailments, but it is highly poisonous, making the cure worse than the disease. • *Hydrocotyles* species have long, creeping stems that form dense mats in aquatic habitats. **Where found:** damp, shady sites, marshes, ponds and swamps; throughout Atlantic Canada. **Also known as:** navelwort.

Angelica

Angelica atropurpurea

Height: to 2.5 m
Leaves: compound with 3 major divisions, pinnately divided into sharp-toothed leaflets, 3–10 cm long
Flowers: small, greenish white, in large, ball-like clusters
Fruit: flat, winged, ribbed seed-like capsules (schizocarps), 3–6 mm long

Angelica in bloom, with its large, spherical clusters of whitish flowers, suggests a display of exploding fireworks. • The leaves smell like parsley and have a strong but pleasant taste. They were traditionally used in soups and stews or to flavour gin and liqueurs. Teas and extracts made from the roots and seeds aid digestion and relieve nausea and cramps. **Where found:** wet, open sites; throughout Atlantic Canada.

Indian-hemp

Apocynum cannabinum

Height: to 1 m
Leaves: 5–10 cm long, opposite, oval, often hairy beneath
Flowers: 2–4 mm long, greenish white, bell-shaped, 5 petals, in open clusters
Fruit: pairs of slender pods (follicles), 10 cm long

This common, weedy native plant is tough and can even push through asphalt. • The stem's long, strong fibres make a durable, fine thread. Traditionally, mature stems were soaked in water to remove the coarse outer fibres, then rolled against the leg into thread. • This plant is poisonous. The milky sap can cause skin blistering; ingestion may result in sickness and death. **Where found:** fields, roadsides, woodland edges and open sites; absent from Labrador and PEI. **Also known as:** dogbane.

Common Milkweed

Asclepias syriaca

Height: 60–180 cm
Leaves: 10–25 cm long, opposite, oblong, hairy beneath
Flowers: tiny, pinkish lavender, fragrant, numerous, in rounded clusters
Fruit: spiny follicles in erect clusters

By far our most common milkweed, these weedy plants contain glycosides that are toxic to both animals and humans. The insects adapted to feed on these plants become poisonous and tend to be brightly coloured and conspicuous, advertising their toxicity. Monarch butterfly larvae feed solely on milkweed leaves. They absorb the glycosides into their bodies, so both larvae and adult butterflies become poisonous to predators. **Where found:** open sites such as fields, meadows and roadsides; throughout the Maritimes.

Bittersweet Nightshade

Solanum dulcamara

Height: climbing to 2.5 m
Leaves: 5–10 cm long, alternate, oval, 2 basal lobes
Flowers: 2 cm wide, white to blue-violet, 5 backward-curved petals, protruding yellow centre, in drooping clusters
Fruit: red, oblong berries, about 1 cm long

Bittersweet nightshade's immature green berries and leaves contain toxic alkaloids that can cause vomiting, dizziness, convulsions, paralysis and even death. • Extracts of this plant are reported to have antibiotic activity, which could be useful in salves and lotions for combating infection. It also contains beta-solanine, a tumour inhibitor that may have potential for treating cancer. **Where found:** moist sites, open woods, thickets, clearings and disturbed sites; throughout Atlantic Canada. **Also known as:** climbing nightshade.

Phlox

Phlox spp.

Height: 60–120 cm
Leaves: 7.5–15 cm long, mostly opposite, oblong, pointed at tip, veined
Flowers: 2.5 cm across, variable colour, fragrant, funnel-shaped, 5 petals, in clusters
Fruit: 3-valved capsules, 6 mm long

Phlox's sweet-smelling, white, pink, lavender or bluish flowers in a pinwheel-like fan of petals have been cultivated by horticulturalists for years. The flowers have long corolla tubes, which butterflies and moths with long tongues are perfectly adapted to pollinate. • Some phlox species are mat-like, but *P. paniculata* grows upright. **Where found:** open, rocky outcrops and open forests; throughout Atlantic Canada. **Also known as:** summer phlox, garden phlox.

Oysterleaf

Mertensia maritima

Height: 10–35 cm
Leaves: 10–25 mm long, succulent, pale blue-green, oval or nearly round
Flowers: 4–10 mm across, blue (pink in bud), bell-shaped, 5 lobed, in small clusters on stems above the leaves
Fruit: 4-chambered capsules (schizocarps),

Oysterleaf is a plant of beaches and coastal areas. It is easily recognized by its blue flowers and, when not in bloom, by its succulent, blue-green leaves. • This member of the borage family is considered edible, and most members of this family have seeds that contain interesting fatty acids. • The genus name *Mertensia* honours Franz Karl Mertens (1764–1831), a renowned German botanist who was primarily a collector of algae. **Where found:** beaches; from NL south. **Also known as:** sea-lungwort, sea bluebells.

Blue Vervain

Verbena hastata

Height: 40–150 cm
Leaves: 4–18 cm long, opposite, lance-shaped, coarsely toothed
Flowers: 6 mm wide, violet-blue, 5 petals, densely clustered in a blunt-tipped spike
Fruit: 2-parted capsules

An abundant and conspicuous denizen of damp, open ground, blue vervain produces spikes of showy blossoms that bloom in spirals, working their way up the flower spike. This pretty wetland plant is important for insects, attracting numerous bees and butterflies. • The steamed leaves are palatable, and the flowers make a pretty garnish or addition to salads. **Where found:** marshes, ditches, wet meadows and shorelines; throughout Atlantic Canada but absent from PEI.

Heal-all

Prunella vulgaris

Height: 10–50 cm
Leaves: 2.5–7.5 cm long, opposite, lance-shaped, finely toothed
Flowers: 1–2 cm long, purplish to pink, tubular, hood-like upper lip arches over 3-lobed lower lip (middle lobe fringed), in dense, spike-like clusters
Fruit: dark, shiny, ribbed nutlets

Introduced from Eurasia, this abundant and often weedy little mint ranges throughout North America. As its name suggests, it has been used to treat just about every type of ailment. Research indicates that the entire plant has antibacterial compounds and inhibits the growth of a number of disease-causing bacteria. **Where found:** open, weedy areas, lawns, fields and roadsides; throughout Atlantic Canada. **Also known as:** self-heal.

Wood Sage

Teucrium canadense

Height: to 1 m
Leaves: 5–13 cm long, opposite, lance-shaped, toothed, deeply veined
Flowers: 7.5–20 mm long, scentless, white, pink or lavender, 2-lipped, the lower lip much larger, in a crowded, spike-like cluster
Fruit: yellowish brown, ellipsoid nutlets, 2 mm long

This member of the mint family has an unpleasant taste, and for perhaps this reason alone, it was thought to have medicinal properties (since culinary ones were lacking). • Teucer, the first king of Troy, was reputedly the first to use this plant medicinally. It has been used to heal sores and treat ulcers, and is sometimes sold commercially as "pink skullcap." **Where found:** thickets, woods and shorelines; from NB and NS south. **Also known as:** Canada germander.

Common Monkeyflower

Mimulus ringens

Height: 60–120 cm
Leaves: 5–10 cm long, opposite, lance-shaped, clasping stem
Flowers: 2–3 cm long, lavender, 2-lipped corolla, yellow "throat," on long stalks
Fruit: oblong capsules with many yellow seeds

When looked at head on, this flower resembles a monkey's face, hence the plant's common name. • This relative of garden snapdragons brightens moist, open habitats with its pale purple blooms. After the flowers drop, the oblong fruit capsules appear, formed by the sepals fusing into inflated balloons full of seeds. **Where found:** marshes, wetlands and damp, open sites; throughout Atlantic Canada.

Wood-betony

Pedicularis canadensis

Height: 15–45 cm
Leaves: 7–13 cm long, basal or alternate on the stem, elliptical, fern-like
Flowers: 2 cm long, pale yellow, maroon or bicoloured, 2-lipped with a hood-like upper lobe, leaf-like bracts, in a spike
Fruit: dry capsules, 1–2 mm long

This odd-looking plant depends on mycorrhizal fungi for nutrient intake and should not be transplanted or disturbed because it will not survive. • The genus name derives from *pediculus*, meaning "louse." It was once thought that if cattle consumed wood-betony, they would become louse-ridden. **Where found:** moist to dry, forested habitats; throughout QC, the Gaspé Peninsula and the St. Lawrence region; rare in NB. **Also known as:** lousewort.

Common Bladderwort

Utricularia vulgaris ssp. macrorhiza

Height: 10–25 cm, aquatic
Leaves: 0.6–2 cm long, usually underwater, alternate, divided into thread-like segments, interspersed with tiny bladders, 2 mm across
Flowers: 1–2 cm long, bright yellow with some red spotting, resembles a snapdragon, in a raceme of 6–20 above the water
Fruit: tiny, single-chambered capsules

Like other carnivorous plants, bladderworts are typically found in cold, acidic, nitrogen-poor environments. They get their nitrogen from the invertebrates they digest, so they are able to grow where other plants cannot survive. • The bladders remain deflated until their trigger hairs are disturbed, then the tiny pouches open abruptly, sucking in water and the hapless aquatic creatures that set off the trap. **Where found:** ponds, lakes and marshes; throughout Atlantic Canada. **Also known as:** *Utricularia macrorhiza*.

Harebell

Campanula rotundifolia

Height: 15–50 cm
Leaves: 1.5–8 cm long, basal leaves oval to heart-shaped and toothed, stem leaves alternate and narrowly lance-shaped
Flowers: 1.5–3 cm long, purple-blue, 5-lobed, bell-shaped, nodding
Fruit: nodding, oblong to cone-shaped capsules

The genus name *Campanula* means "little bell" and describes the nodding flowers. However, not all *Campanula* species are characteristic, and some in the east have flared petals. • The common garden bellflower (*C. rapunculoides*) frequently escapes from gardens and shows up in natural areas; it differs by having heart-shaped leaves. • Victorians attributed the sentiments of humility or grief to harebell with its downcast face. **Where found:** rocky banks and slopes, meadows and shorelines; throughout Atlantic Canada. **Also known as:** bluebell.

Cardinal-flower

Lobelia cardinalis

Height: 60–120 cm (occasionally reaches 2 m)
Leaves: 5–15 cm long, alternate, lance-shaped, toothed
Flowers: 2–4.5 cm long, intense red or scarlet, 2-lipped, in showy spikes
Fruit: spherical capsules with brown seeds

Cardinal-flower gets its name from the striking colour of its flowers, which matches that of the robes of cardinals in the Roman Catholic Church. The brilliant red blossoms with their long corolla tubes are well adapted to attract this plant's primary pollinators, hummingbirds. • Some *Lobelia* species, including cardinal-flower, have therapeutic properties, but large doses can cause nausea, vomiting, drowsiness, respiratory failure and potentially death. **Where found:** wet woods, meadows and damp areas; from QC to NB.

Bluets

Houstonia caerulea

Height: 15 cm
Leaves: 0.5–1.25 cm long, mostly basal, stem leaves opposite, spatulate
Flowers: 1.25 cm wide, tubular, light blue to white, yellow centre, 4 petals, solitary on long stalks
Fruit: flattened capsules, 3 mm long

Bluets grow in masses, carpeting the ground in spring. The small flowers' bright yellow centre acts like a bull's-eye for insects, guiding tiny bees to the pollen found inside the tubular flowers. • Bluets cannot withstand much competition from other plants and typically grow in barren soil. **Where found:** dry, barren soil in clearings, deciduous woodlands and transitional forests; from QC to NS and south.

Partridgeberry

Mitchella repens

Height: 10–30 cm, trailing along the ground
Leaves: 1.3–2 cm long, opposite, oval to round, leathery
Flowers: 1.3–1.6 cm long, white, fragrant, trumpet-shaped, usually
4 petals, fuzzy inside
Fruit: scarlet berries, 1 cm wide

The reddish berries of this plant are edible, though not very tasty. The Mi'kmaq, Iroquois, Montagnais and Maliseet peoples ate them fresh or cooked into a jam. Wildlife such as ruffed grouse and wild turkeys eat the buds, leaves, flowers and fruit of this plant. • The trailing stems and berries make good Christmas decorations, though taking plants from the wild is discouraged. **Where found:** dry or moist woodlands, rich, mixed forests with acidic soils (especially under conifers) and transitional forests; throughout Atlantic Canada.

Northern Bedstraw

Galium boreale

Height: 20–100 cm
Leaves: 2–6 cm long, in whorls of 4, stalkless, narrowly lance-shaped
Flowers: 3–7 mm wide, white, showy, 4 petals, in clusters at stem tips
Fruit: pairs of hairy nutlets, 2 mm long

Bedstraw is related to coffee (*Coffea* spp.), and its tiny, paired, short-hairy nutlets can be dried, roasted and ground as a coffee substitute. Juice or tea made from bedstraw was used traditionally to treat a variety of skin problems. • Sweet-scented bedstraw (*G. triflorum*) is sweeter smelling and has bristle-tipped leaves and stems. It is also found throughout Atlantic Canada. **Where found:** open sites, roadsides, meadows, open woodlands and shorelines; throughout Atlantic Canada.

Devil's Beggarticks

Bidens frondosa

Height: 60–120 cm
Leaves: to 10 cm long, compound with 3–5 serrated, sharply pointed leaflets
Flowers: ray flowers absent; disc flowers minute, yellow; flowerheads 2–2.5 cm
across, usually solitary
Fruit: cylindrical to ovoid achenes with 2-barbed horns

The barbed fruit of devil's beggarticks easily attach to clothing and animals but are difficult to brush off. This strategy for seed dispersal is annoying, particularly for hikers. • The "tick" part of the common name refers to the fact that, when stuck to clothing, the seeds can look like ticks, which are far more undesirable to hikers than burs. **Where found:** damp soils along streams, lakes, bogs and marshes; throughout Atlantic Canada. **Also known as:** stick tight, bur marigold, devil's pitchfork.

Giant Ragweed

Ambrosia trifida

Height: 1–3.6 m
Leaves: 5–10 cm long, opposite on lower stem, alternate on upper stem, deeply pinnately divided into 3–5 lobes
Flowers: ray flowers absent; disc flowers 6–13 mm long, greenish; flowerheads in terminal spikes to 15 cm tall
Fruit: reddish brown, beaked achenes, 3–4 mm long

Huge stands of this abundant plant sometimes cover old fields, and its airborne pollen causes much hayfever suffering. • Ragweed seeds 4 to 5 times larger than today's wild species have been found in archaeological digs in North America, suggesting that this plant was once selectively bred, perhaps for its seed oil. **Where found:** moist soils, disturbed sites in old fields and along roadsides; QC, NB, NS and PEI.

Common Yarrow

Achillea millefolium

Height: 10–80 cm
Leaves: 5–15 cm long, 1–2 mm wide, basal and alternate on the stem, fern-like
Flowers: ray flowers white (rarely pinkish), usually 5, 3-toothed; disc flowers whitish, minute; flowerheads 5–10 mm wide, in dense, flat-topped clusters
Fruit: flattened achenes, less than 1 cm long, with a single seed

This hardy, aromatic perennial has served for thousands of years as a fumigant, insecticide and medicine. The Greek hero Achilles, for whom the genus was named, reportedly used it to heal his soldiers' wounds after battle. • Yarrow is an attractive ornamental, but its extensive underground stems (rhizomes) are invasive. **Where found:** dry to moist, open sites; throughout Atlantic Canada.

Oxeye Sunflower

Heliopsis helianthoides

Height: 60–150 cm
Leaves: 5–15 cm long, opposite, lance-shaped, serrated
Flowers: rays flowers golden yellow, 8–16; disc flowers minute, yellow; flowerheads to 7.5 cm wide, solitary to several on long stalks
Fruit: 4-angled, slightly hairy achenes

This plant is a member of the sunflower/aster family, as are daisies, asters and the sunflowers (*Helianthus* spp.) that we are familiar with for producing giant heads of "spits" from the disc flowers (the ray flowers do not produce seeds). • Both the ray and disc flowers of oxeye sunflower produce fruits, but it is mainly American goldfinches that feast upon the tiny seeds. **Where found:** semi-shaded woodland borders; throughout Atlantic Canada. **Also known as:** false sunflower.

Common Tansy

Tanacetum vulgare

Height: 40–150 cm
Leaves: 5–25 cm long, alternate, finely divided into narrow, toothed leaflets
Flowers: ray flowers absent; disc flowers tiny, yellow; flowerheads 0.5–1 cm across, button-like, in clusters of 20–200
Fruit: tiny achenes

Common tansy contains toxic volatile oils that are potentially fatal. These plants were traditionally strewn on floors, under mattresses and within stored bed linens and clothing to repel insects, or else they were boiled into an insecticide or a wash to treat lice or scabies infestations. Tansy is also used as a flavouring agent in alcoholic beverages, but never drink tansy tea. **Where found:** disturbed sites and open areas; throughout Atlantic Canada.

Golden Ragwort

Packera aurea

Height: 30–76 cm
Leaves: to 13 cm long, basal leaves heart-shaped, purplish beneath, with scalloped edges, stem leaves alternate and lobed
Flowers: ray flowers yellow, 10–12, 1 cm long; disc yellow, 1 cm wide; flowerheads several on long stalks in open clusters
Fruit: brownish, cylindrical achenes with tufted, white pappus

In early spring, colonies of golden ragwort brighten shady road-sides and low-lying meadows. The basal leaves are dark purple below, possibly allowing the plant to absorb additional heat to aid growth during cool spring weather. • Historically, ragwort tea was used medicinally, but this plant contains toxic alkaloids and should not be consumed. **Where found:** moist woods and fields; throughout Atlantic Canada.

Canada Goldenrod

Solidago canadensis

Height: 60–150 cm
Leaves: 7–15 cm long, narrowly lance-shaped, stalkless, rough-hairy, sawtoothed
Flowers: ray flowers yellow, 10–17, 1–3 mm long; disc flowers yellow, 6–12; flowerheads numerous in plume-like clusters
Fruit: hairy achenes with tufted, white pappus

Canada goldenrod is a classic pioneer forb of old fields. Growth-inhibiting enzymes released from its roots discourage other plants from flourishing. • Many people think that goldenrod flowers cause hayfever, but the real culprit is probably ragweed, which shares the same habitat. • The goldenrod gall fly (*Eurosta solidaginis*) lays its egg in the stem, which causes the plant to form a hardened, perfectly round mass of tissue that looks like it swallowed a golf ball. **Where found:** moist to dry fields and open sites; throughout Atlantic Canada.

New England Aster

Symphyotrichum novae-angliae

Height: 60–150 cm
Leaves: 2.5–10 cm long, lance-shaped, clasping the stem
Flowers: ray flowers violet, rose or magenta, 45–100, 1 cm long; disc flowers yellow-orange, 1 cm wide; flowerheads showy, in leafy clusters
Fruit: hairy achenes with tufted, bristly pappus

Of the many asters in our region, this one is the showiest. Though considered an aggressive weed, New England aster is admired for its attractive flowers, and it is often deliberately planted in flowerbeds. The flowers are typically rich purplish magenta but can vary to lilac and almost white, and the blooms persist after the first autumn frosts. **Where found:** moist to dry meadows, wet thickets, swamps and open sites; throughout Atlantic Canada.

Common Boneset

Eupatorium perfoliatum

Height: 60–120 cm
Leaves: 7–20 cm long, opposite, fused at base, lance-shaped, sparsely hairy, serrated
Flowers: ray flowers absent; disc flowers white, 9–23; flowerheads in flat-topped clusters
Fruit: resinous achenes with tufted, white pappus

Early herbalists believed the perforated leaves of this wetland plant indicated that it was useful in setting bones, so the leaves were wrapped with bandages around splints. There is little scientific proof that the leaves help bones to heal, but researchers in Germany have found compelling evidence that this plant boosts the immune system. **Where found:** wet meadows and low, moist sites; throughout Atlantic Canada.

Canada Thistle

Cirsium arvense

Height: 60–150 cm
Leaves: 5–15 cm long, alternate, usually stalkless, deeply lobed, spiny-toothed, wavy edges
Flowers: ray flowers absent; disc flowers numerous, tubular, purple or pink; flowerheads 1.2–2.5 cm across
Fruit: seed-like achenes

Introduced to Canada from Scotland in the 17th century, this thistle is considered a noxious weed, choking out other plants and reducing crop yields. Its deep underground runners contain a substance that inhibits the growth of nearby plants. Each year, one plant can send out up to 6 m of runners, and female plants can release up to 40,000 seeds. • This thistle is edible, prepared and eaten much like an artichoke, to which it is related. **Where found:** disturbed sites; throughout Atlantic Canada.

Common Dandelion

Taraxacum officinale

Height: 5–50 cm
Leaves: 5–40 cm long, basal rosette, oblanceolate, lobed
Flowers: ray flowers numerous, yellow; disc flowers absent; flowerheads 2.5–5 cm wide, solitary on hollow stalks
Fruit: tiny achenes with fluffy, white pappus

Emerald green lawns sprinkled with yellow dandelion blossoms create a rather showy palette but rankle fastidious lawn-keepers. • Brought to North America from Eurasia, dandelions were cultivated for food and medicine. Young dandelion leaves and flowerheads are full of vitamins and minerals and make nutritious salad greens. The roots can be ground into a caffeine-free coffee substitute or boiled to make a red dye. **Where found:** disturbed sites; throughout Atlantic Canada.

Broad-leaved Arrowhead

Sagittaria latifolia

Height: to 1 m, aquatic
Leaves: 15–40 cm long, basal, arrowhead-shaped, long-stalked
Flowers: 2 cm long, white, 3 petals, in 2–10 whorls of 3 flowers on erect stalks
Fruit: beaked achenes in clustered heads

The leaves of this characteristic marsh plant can vary in shape but always have long basal lobes. • The entire rootstock is edible, but the corms are preferred. When cooked, they taste like potatoes or chestnuts but are unpleasant raw. Native peoples often camped near arrowhead sites for weeks, harvesting the crop or seeking out muskrat caches of corms. **Where found:** shallow water or mud in sunny marshes, ditches and other wetlands; throughout Atlantic Canada. **Also known as:** duck potato.

Skunk Cabbage

Symplocarpus foetidus

Height: 30–60 cm
Leaves: 38–55 cm long, basal, heart-shaped, cabbage-like
Flowers: tiny, yellowish, star-like, clustered on a ball-like spadix inside a brownish purple, hooded spathe, 10–15 cm high
Fruit: brown-black berries in rounded clusters

This odd-looking plant is our first wildflower to bloom each year. As early as late February, the spathes push from the boggy ground, aided by heat produced through cellular respiration that melts nearby snow and ice. The giant, cabbage-like leaves emerge after the flowers have bloomed. • This wetland plant emits a mild odour when left alone but reeks when damaged. The smell attracts pollinating insects but repels animals that may eat or otherwise damage the plant. **Where found:** spring-fed, boggy wetlands; throughout Atlantic Canada.

Jack-in-the-Pulpit

Arisaema triphyllum

Height: 30–90 cm
Leaves: 7.5–15 cm long, basal, 1–2, compound with 3 oval, pointed leaflets
Flowers: tiny, in clusters on a greenish, club-shaped spadix, 7.5 cm long, encircled by a green to purplish, flanged spathe
Fruit: smooth, shiny, red berries

Jack-in-the-pulpit's odd flowers are composed of a spadix ("Jack") covered with tiny male and female flowers and surrounded by a hood-like spathe (the "pulpit"). The spathe colour varies from bright green to boldly purple-striped. • The entire plant and the bright red berries are poisonous if eaten fresh. Native peoples cooked the fleshy taproots as a vegetable or dried and pounded them into flour. **Where found:** moist hardwood forests, damp woods and swamps; from QC to NB and south. **Also known as:** Indian turnip.

Lesser Duckweed

Lemna minor

Height: tiny, in colonies
Leaves: 2–5 mm across, oval
Flowers: minute, usually 3, without sepals or petals
Fruit: tiny, thin-walled, bladder-like utricles

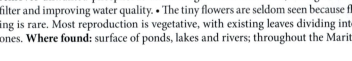

Lesser duckweed is the most common of a number of tiny, floating, aquatic plants that are often mistakenly called algae. Sometimes regarded as a nuisance, duckweed removes unwanted phosphorus and nitrogen from the water, acting as a natural filter and improving water quality. • The tiny flowers are seldom seen because flowering is rare. Most reproduction is vegetative, with existing leaves dividing into new ones. **Where found:** surface of ponds, lakes and rivers; throughout the Maritimes.

Pickerelweed

Pontederia cordata

Height: 90 cm
Leaves: 10–20 cm long, mostly basal, variable heart-, arrow- or lance-shaped, shiny, many parallel veins
Flowers: 2.5 cm wide, lavender to white, funnel-shaped, 6 petals, uppermost petal marked with yellow, in spikes on stalks 60–90 cm tall
Fruit: dry, ridged achenes

This beautiful wetland species is common in many marshland areas, but it has not fared well in competition with recent non-native invading plants. • The young leaves may be eaten as a salad green or potherb, cooked like spinach or added to soups. The starchy seeds have a nutty flavour and can be eaten fresh, dried or roasted and ground into a powder. **Where found:** shallow water in marshes, lakes, streams and ponds; QC and the Maritimes.

Broad-leaved Cattail

Typha latifolia

Height: 1–3 m
Leaves: up to 3 m long, 1–2 cm wide, flat, grass-like
Flowers: tiny, numerous, in dense spikes 15–20 cm long; female flowers dark brown, on lower portion of spike; male flowers cone-shaped, on upper portion of spike, disintegrating and leaving stem tip bare
Fruit: tiny, ellipsoid achenes clustered in a fuzzy, brown spike (cattail)

Cattails rim wetlands and line lakeshores and ditches across North America, providing cover for many animals and critical habitat for marsh birds. • This plant's long rhizomes were traditionally eaten fresh in spring. Later in the season, they were peeled and roasted or dried and ground into flour. • Narrow-leaved cattail (*T. angustifolia*) is more southern ranging and has narrower leaves and a gap between the male and female flowers on the spike. **Where found:** wetlands, marshes, ponds, ditches and damp ground; throughout Atlantic Canada.

Beach Grass

Ammophila breviligulata

Height: up to 1 m
Leaves: 1–2 cm wide, up to 40 cm long, deeply veined above, smooth below
Flowers: tiny, inconspicuous, in spike-like panicles to 25 cm long
Fruit: minute seeds

Beach grass is a very important pioneer plant that stabilizes sand dunes. The stems continue to grow higher as blowing sand builds up around them. The buried stems eventually become roots that reach and spread far and wide under the sand. The flower spikes produce seeds that spread with wind. • The tough, thin, green blades roll up to conserve moisture on hot days and unroll to catch moisture. **Where found:** beaches and sand dunes close to the sea; throughout Atlantic Canada.

Common Reed Grass

Phragmites australis

Height: 1–3 m
Leaves: 20–40 cm long, 1–4 cm wide, flat, stiff
Flowers: tiny, purple-brown, in plume-like panicles 10–40 cm long
Fruit: seed-like grains

This giant non-native grass is conspicuous, forming massive stands that tower over other plants. • The stem contains a sugary substance with many uses. Dried, ground stalks can be turned into a sugary flour that will bubble and brown like a marshmallow when heated, and sugar crystals from dried stems or the sweet, gummy substance that bleeds from cut stems can be eaten like candy. **Where found:** marshes, ditches, wetlands and shorelines; throughout Atlantic Canada.

Rush

Juncus spp.

Height: to 1.5 m
Leaves: basal, 1–2 mm wide, flattened or round
Flowers: small, greenish to brown, usually 3 petals and 3 sepals, numerous in open clusters or dense, terminal heads
Fruit: brown capsules with many tiny seeds

Rushes are a type of grass that stabilizes dunes and provides shelter for a variety of wildlife. Few other plants can tolerate tidal environments that transition between terrestrial and underwater. These plants occur around the world, mostly in temperate zones. **Where found:** salt marshes, coastal and freshwater beaches, estuaries, wetlands, lakes, riparian zones and rural or disturbed sites; throughout Atlantic Canada.

Cord Grass

Spartina patens

Height: 15–80 cm
Leaves: 10–50 cm long, 1–4 mm wide, light green, rolled inward
Flowers: inconspicuous, in heads 5–22 cm long, with 1–4 spikes, each 1–8 cm long
Fruit: wheat-like seeds

Dense mats of cord grass appear tousled and messy, and each year, new grass adds density to the previous year's growth, producing increasingly thick and dense mats. Cord grass is tenacious, growing from rhizomes as well as seeds, and it is able to grow in salt water because root membranes screen out much of the salt. It is an important component of a healthy saltmarsh ecosystem. **Where found:** salt marshes along the coast; throughout Atlantic Canada. **Also known as:** saltmarsh hay.

Spike Grass

Distichlis spicata

Height: 10–50 cm
Leaves: 2–10 cm long, narrow, flat, stiff
Flowers: inconspicuous, clustered in greenish to purple spikelets, to 8 cm long
Fruit: tiny, smooth, oval seeds

This coastal grass is typical of salt marshes but also occurs in a variety of habitats, including forests, desert-scrub and montane zones. It has a heavy root system that develops into a thick sod layer over time. • The ability of this plant to excrete salt from its tissues is remarkable, and stems can often be seen covered in salt crystals. It is valuable as a grazing crop for livestock for its ability to use salt water, important in times of drought. **Where found:** salt marshes; throughout the Maritimes. **Also known as:** seashore salt grass.

Beach Sedge

Carex silicea

Height: under 1 m
Leaves: basal, 10–25 cm long, 1–5 mm wide,
Flowers: inconspicuous, clustered in silvery green spikes,
4–8 cm long, arching or arching or nodding
Fruit: tiny, elliptical achenes

Sedges are grass-like herbs, and the genus name *Carex* is a Latin word that translates as "grasses with sharp leaves." However, unlike grasses, the stems of *Carex* species are solid and often triangular in cross-section. • These plants are found around the world, and their edible grains have served as famine foods. Beach sedge is common in our region and, as the name implies, it is largely coastal. **Where found:** beaches; throughout Atlantic Canada.

Cottongrass Bulrush

Scirpus cyperinus

Height: 50–100 cm
Leaves: to 30 cm long, 2–5 mm wide, grass-like, rough
Flowers: inconspicuous, in greenish spikelets in large, arching, terminal
clusters that become fuzzy as they mature
Fruit: sharply beaked, yellowish achenes, 1 mm long

Like other grass-like plants, bulrushes are great for weaving, and many species are edible. The seeds are food for ducks and other marsh birds. The long stems offer cover to birds and other animals, and are eaten by raccoons, muskrats and geese. • Tenacious bulrushes anchor sandy soil and build up land in low-lying flooded areas. **Where found:** open, wet places and shorelines; throughout Atlantic Canada; absent from Labrador, rare in Newfoundland. **Also known as:** wool grass.

Ostrich Fern

Matteuccia struthiopteris

Height: 50–150 cm
Leaves: 2 types of fronds; sterile fronds lance-shaped, to 30 cm wide, tapered
at both ends, with 20–60 leaflet pairs, fronds form a crown-like cluster; fertile
fronds 20–60 cm tall, narrow, erect, olive green becoming dark brown, arising from
the centre of the clump

The emerald green, tightly coiled fiddleheads are a traditional dish in Québec and New Brunswick, a delicious wild edible that is also sold commercially. Fiddleheads are collected in spring when they are about 15 cm tall and still tightly coiled. Do not overharvest from a single plant—leave more than half, or at least 3 tops, or the plant may die. **Where found:** rich, most or wet soils along streams and riverbanks and in open woods and swamps; throughout Atlantic Canada; absent from Labrador.

Bracken Fern

Pteridium aquilinum

Height: fronds to 3 m
Leaves: triangular, hairy, 2–3 times pinnately divided into
10 or more pairs of leaflets

This widespread species is common around the
world. The deep rhizomes spread easily and allow this
plant to survive fires. • Fiddleheads and rhizomes from
this fern were once a popular food item, but they have
proven to carry potent carcinogens and are no longer considered safe to eat. • The fronds are resistant to decay and have been
used historically for such things as thatching and packing material. **Where
found:** fields, meadows, bogs, disturbed sites, clearings, dry to wet forests and
lakeshores; throughout Atlantic Canada.

Sea Lavender

Limonium carolinianum

Height: 10–70 cm
Leaves: 1–30 cm long, spoon-shaped, leathery, in a basal rosette
Flowers: 4–10 mm across, light purple or lavender pink, tubular, 5 petals,
along one side of the stem
Fruit: small capsules with a single seed

This colourful seaside perennial is not related to the herb
lavender; the common name is merely a reference to the flower
colour. Dense colonies of this plant grow in some salt marshes, salt
meadows and dunes. • The rootstock was once made into a mouthwash because
of the astringent tannins, which have been and are still used today for a variety of
home remedies. **Where found:** coastal salt marshes; Newfoundland and QC south
to Florida. **Also known as:** harsh rosemary.

Eel Grass

Zostera marina

Length: up to 3 m
Width: up to 1.3 cm

Neither algae nor a grass but a flowering plant, this species is
a type of sea grass. Eel grass is truly marine, spending almost its
entire life underwater. Rarely, this plant is exposed at low tide, when
its long, narrow, bright green strands can be seen in shallow, rocky
waters. • Sea grass flowers are tiny and inconspicuous because there is
no need to attract insects for pollination. • Eel grass provides habitat and food for
worms, snails, algae, crabs, lobsters, sponges, fish, shellfish and waterfowl. **Where
found:** low intertidal and subtidal zones; throughout Atlantic Canada.

Sugar Kelp – Brown Algae

Laminaria saccharina

Length: up to 3 m

There are 3 types of algae—brown, red and green. Kelp is a brown alga and has root-like anchors that affix to rocks while the wide (15 cm), leathery blades float with the currents. • Forests of kelp create habitat for marine animals such as sea urchins, sea stars, periwinkles, crabs and small fish, and are among the most biodiverse forests, including terrestrial ones. • A sweet-tasting powder forms on the blades when they dry out. Sugar kelp has a nutty flavour and makes a tasty snack when dried. **Where found:** low littoral rock pools; from low-tide line and subtidal zones to below 20 m; throughout Atlantic Canada. **Also known as:** sugar wrack.

Rockweed – Brown Algae

Fucus distichus

Length: to 50 cm

This small, tufted, brown alga lacks the air bladders that are fun to pop on other rockweeds known as bladder wrack (*F. vesiculosus*), though the tips of the fronds do swell when mature. The rockweed family includes many of the most widely distributed seaweeds in the world and gets its name from the habit these plants have of adhering to rocks and other solid foundations. **Where found:** in tidal pools on rocky shores; middle and lower intertidal zones all along the coast.

Encrusting Corralline Algae – Red Algae

Corallina officinalis

Length: 6–12 cm

Red algae are the largest group of seaweeds. A red pigment typically masks the chlorophyll that would otherwise render these algae green in colour. • This red alga looks like coral because of its branching, calcified fronds and its whitish pink to lilac colouration. Most tidal rock pools will have corralline algae, and it provides habitat for many small sea creatures. • Encrusting corralline algae has been used in the research of bone replacement therapies. **Where found:** often forms a distinct zone just below the rim of rock pools, in middle and lower shore zones; throughout the Maritimes. **Also known as:** coral weed.

Dulse – Red Algae

Palmaria palmata

Length: fronds, 5–30 cm

This rubbery alga can be eaten raw like a vegetable and is commercially harvested to be dried and sold in local grocery stores. Dried dulse can be added to soups and to seafood and Asian dishes. It is loaded with vitamins and minerals. • Nori or laver (*Porphyra* spp.), another red alga found in our waters, is cultivated in East Asia and is routinely used in Japanese cooking. • Dulse grows best in deeper waters because the red pigment photosynthesizes better in dim underwater light. **Where found:** on rocks and mussels; epiphytic on several algae; intertidal (particularly near low water) and shallow subtidal zones; throughout Atlantic Canada. **Also known as:** *Ptilota elegans, Rhodymenia palmata.*

Irish Moss – Red Algae

Chondrus crispus

Length: 15 cm

Clumps of this seaweed on the beach or floating in shallow water look like moss and give the species its analogous common name. • The walls of this and similar red algae are made up of cellulose, agars and carrageenans—long-chain polysaccharides—that have widespread commercial uses. Agars are used as thickeners in soups and dairy products and as mediums for growing cultures in laboratories. Carrageenan is used as a thickener in cooking and baking and as an ingredient in cough syrups. **Where found:** on rocks and in pools; lower intertidal and shallow subtidal zones; throughout Atlantic Canada. **Also known as:** carrageenan moss.

Sea Lettuce – Green Algae

Ulva lactuca

Length: tissue thin sheets, 18–30 cm long

Among the few green algae species found in the intertidal zone, perhaps the most visible and abundant is the bright green sea lettuce, either attached to rocks or free-floating. This seaweed has a very simple structure and is only 1 or 2 cells thick. • Sea lettuce is of high caloric value and is eaten by crabs and molluscs. Although it looks like salad lettuce, it is not tasty. • Sea lettuce is tolerant of fresh water or minor pollution for short periods. **Where found:** shallow bays, lagoons, harbours and marshes, on rocks and on other algae; intertidal and high tide zones; throughout Atlantic Canada.

REFERENCES

AmphibiaWeb. amphibiaweb.org.

Bird Studies Canada. www.bsc-eoc.org.

Burrows Roger. 2002. *Birds of Atlantic Canada*. Lone Pine Publishing, Edmonton.

Canadian Amphibian and Reptile Conservation Network (CARCNET). www.carcnet.ca.

Canadian Biodiversity. canadianbiodiversity.mcgill.ca.

Canadian Wildlife Federation. www.cwf-fcf.org.

Centre for Marine Biodiversity. www.marinebiodiversity.ca.

Committee on the Status of Endangered Wildlife in Canada (COSEWIC). www.cosewic.gc.ca.

Eder, Tamara. 2011.*Whales and Other Marine Mammals of Atlantic Canada*. Lone Pine Publishing, Edmonton.

Eder, Tamara, and Gregory Kennedy. 2010. *Mammals of Canada*. Lone Pine Publishing, Edmonton.

Environment Canada. www.ec.gc.ca.

Fisher, Chris, Amanda Joynt, and Ronald J. Brooks. 2007. *Reptiles and Amphibians of Canada*. Lone Pine Publishing, Edmonton.

Gibson, Merritt, and Soren Bondrup-Nielsen. 2008. *Winter Nature: Common Mammals, Birds, Trees and Shrubs of the Maritime Provinces*. Gaspereau Press, Kentville.

Leatherwood, Stephen, and Randall R. Reeves. 1983. *The Sierra Club Handbook of Whales and Dolphins*. Sierra Club Books, San Francisco.

MacKinnon, Andy, et al. 2009. *Edible and Medicinal Plants of Canada*. Lone Pine Publishing, Edmonton.

National Audubon Society. 1998. *Field Guide to North American Fishes, Whales and Dolphins*. Chanticleer Press, Toronto.

National Audubon Society. 1998. *Field Guide to North American Seashore Creatures*. Chanticleer Press, Toronto.

Sheldon, Ian, and Tamara Eder. 2000. *Animal Tracks of Atlantic Canada*. Lone Pine Publishing, Edmonton.

Sibley, D.A. 2000. *National Audubon Society: The Sibley Guide to Birds*. Alfred A. Knopf, New York.

Sibley, David Allen. 2009. *The Sibley Guide to Trees*. Alfred A. Knopf, New York.

Sutton, Ann, and Myron Sutton. 1985. *Eastern Forests*. Audubon Society Nature Guides. Alfred A. Knopf, New York.

GLOSSARY

achene: a seed-like fruit (e.g., sunflower seed)

alcid: a seabird of the family Alcidae, which includes murres, auklets and puffins, characterized by wings well suited to diving and swimming and that beat rapidly in flight

alga (*pl.* algae): simple photosynthetic aquatic plants lacking true stems, roots, leaves and flowers, and ranging in size from single-celled forms to giant kelp

alternate: an arrangement of leaves along a stem in which the leaves are unpaired

anadromous: fish that migrate from salt water to fresh water to spawn (e.g., salmon)

annual: a plant that lives for only one year or growing season

aquatic: water frequenting

arboreal: tree frequenting

Arctic: the region north of the Arctic Circle (66°33' N latitude)

aril: an extra or specialized seed covering, often fleshy, hairy or brightly coloured (e.g., the red, translucent flesh that surrounds a pomegranate seed)

arthropod: an invertebrate with a heard, segmented exoskeleton and paired, jointed legs (e.g., insects, spiders, crustaceans)

awn: a stiff, bristle-like projection, especially from the seed of a grass or grain

axil: the point at which a leaf attaches to a stem

baleen: strands of keratin that hang in sheets from the upper jaws of some whales and are used to filter food from water

barbels: fleshy, whisker-like appendages found on some fish

barren: a mainly treeless area with low-growing plants tolerant of exposure and low-nutrient soils; also known as a "heath" for the dominant plant family found there

basal leaf: a leaf arising from the base of a plant

benthic: living at or near the bottom of a sea or lake

benthos: the community of organisms (plants and animals) that lives at or near the bottom of a lake or sea

berry: a fleshy fruit, usually with numerous seeds

bog: a wetland with poor drainage and a thick layer of peat that only receives nutrients from rainfall; mosses, particularly sphagnum moss, are common vegetation in bogs; *see also* **fen, string bog**

bow riding: a behaviour seen in dolphins in which the animals swim in the bow waves of boats

bract: a leaf-like structure arising from the base of a flower or inflorescence

breach: a whale display in which the animal rises vertically into the air, clearing the water's surface with almost all of its body before splashing back in

brood parasite: a bird that parasitizes other bird's nests by laying its eggs and then abandoning them for the parasitized birds to raise (e.g., brown-headed cowbird)

bulb: a fleshy underground organ with overlapping, swollen scales (e.g., onion)

calyx: a collective term for the sepals of a flower

cambium: inner layers of tissue that transport nutrients up and down a plant stalk or trunk

canopy: the fairly continuous cover provided by the branches and leaves of adjacent trees

capsule: a dry fruit that splits open to release seeds

carapace: a protective bony shell (e.g., of a turtle) or exoskeleton (e.g., of a beetle)

carnivorous: feeding primarily on meat

carrion: decomposing animal matter or a carcass

catadromous: fish that migrate from fresh water to the sea to spawn (e.g., eels)

catkin: a spike or hanging cluster of small flowers

cetacean: a marine mammal of the order Cetacea, which includes whales, dolphins and porpoises

compound leaf: a leaf separated into 2 or more divisions called leaflets

cone: the fruit produced by a coniferous plant, composed of overlapping scales around a central axis

coniferous: cone bearing; seed (female) and pollen (male) cones are borne on the same tree in different locations

corm: a swollen underground stem base that resembles a bulb and is used by some plants as an organ of propagation

corvid: a member of the family Corvidae, which includes crows, jays, magpies and ravens

crepuscular: active primarily at dusk and dawn

cryptic colouration: a colouration pattern designed to conceal an animal

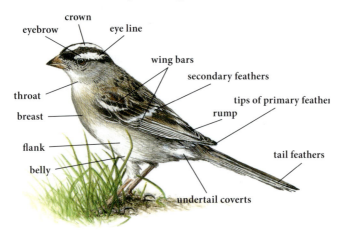

deciduous: a tree whose leaves turn colour and are shed annually; can also refer to petals that fall after a plant flowers

defoliate: to drop leaves

disc flower: a small flower in the centre, or disc, of a composite flower (e.g., aster, daisy or sunflower)

diurnal: active primarily during the day

dorsal: the top or back

drupe: a fleshy fruit with a stony pit (e.g., peach, cherry)

echolocation: navigation by rebounding sound waves off objects to target or avoid them

ecological niche: an ecological role filled by a species

ecoregion: a geographical region distinguished by its geology, climate, biodiversity, elevation and soil composition

ectotherm: an animal that regulates its body temperature behaviourally from external sources of heat, i.e., from the sun

eft: the terrestrial juvenile stage of a newt

elytra (*sing.* elytron): the hardened forewings of some insects, especially beetles, that serve as protection for the underlying hindwings, which are used for flight

endotherm: an animal that regulates its body temperature internally

epiphyte: a plant that grows on another plant but is not a parasite (e.g., many mosses and lichens)

estivate: a state of inactivity and a slowing of the metabolism to permit survival during extended periods of high temperatures and inadequate water supply

estuary: a partly enclosed coastal body of water where fresh water from one or more rivers mingles with salt water from an ocean or sea

eutrophic: a nutrient-rich body of water with an abundance of algae and a low level of dissolved oxygen

evergreen: having green leaves through winter; not deciduous

exoskeleton: a hard outer encasement that provides protection and points of attachment for muscles (e.g., on insects and crustaceans)

fen: a wetland with poor drainage and a thick layer of peat into which water seeps from nearby soils, making the fen's soil more nutrient rich than that of a bog; primary vegetation includes mosses and some grass species; *see also* **bog**

flight membrane: a membrane between the fore and hind limbs of bats and some squirrels that allows these animals to glide through the air; *see also* **patagium**

fluke: *n.* either of the lobes of a whale's tail; *v.* a whale behaviour in which the animal raises its tail above the water before diving

follicle: the structure in the skin from which hair or feathers grow; a dry fruit that splits open along a single line on one side when ripe; a cocoon

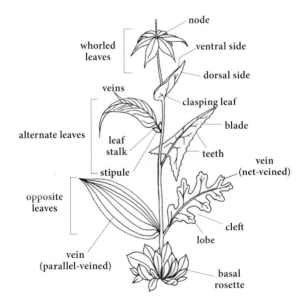

food web: the elaborate, interconnected feeding relationships of living organisms in an ecosystem

forb: a broad-leaved plant that lacks a permanent woody stem and loses its aboveground growth each year; may be annual, biennial or perennial

fry: the young of fish

gillrakers: long, thin, fleshy projections that protect delicate gill tissue from particles in the water

glandular: similar to or containing glands

glaucous: covered with a waxy, whitish coating or "bloom"

graminoid: a herbaceous plant with narrow leaves growing from the base; includes grasses and grass-like plants such as sedges and rushes

habitat: the physical area in which an organism lives

hawking: feeding behaviour in which a bird leaves a perch, snatches its prey in midair and returns to the perch

haw: the small, berry-like fruit of a hawthorn

herbaceous: feeding primarily on vegetation

hibernaculum (*pl.* hibernacula): a shelter in which an animal, usually a mammal, reptile or insect, chooses to hibernate

hibernation: a state of decreased metabolism and body temperature and slowed heart and respiratory rates to permit survival during long periods of cold temperatures and diminished food supply

hip: the berry-like fruit of some plants in the rose family (Rosaceae)

hybrid: the offspring from a cross between parents belonging to different varieties or subspecies, and sometimes between different subspecies or genera

incubate: to keep eggs at a relatively constant temperature until they hatch

inflorescence: a cluster of flowers on a stalk

insectivore: feeding primarily on insects

invertebrate: any animal lacking a backbone (e.g., worms, slugs, crayfish, shrimp)

involucral bract: one of several bracts that form a whorl below a flower or flower cluster

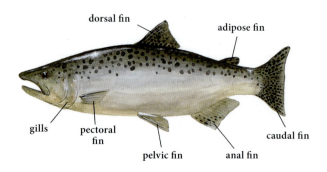

dorsal fin

adipose fin

gills

pectoral fin

pelvic fin

anal fin

caudal fin

key: a winged fruit, usually of an ash or maple tree; also called a **samara**

kype: the hook-like protrusions that develop on the mandibles of a spawning trout or salmon

larva (*pl.* larvae): the immature form of an animal that differs from the adult

leaflet: a division of a compound leaf

lenticel: a slightly raised portion of bark where the cells are packed more loosely, allowing for gas exchange with the atmosphere

leveret: a young hare

littoral: along the shore of a river, lake or sea

lobate: having each toe individually webbed

lobe: a projecting part of a leaf or flower, usually rounded

lobtail: a display in which a whale forcefully slaps its tail flukes on the water's surface

metabolic rate: the rate of chemical processes in an organism

metamorphosis: the developmental transformation of an animal from a larval stage to a sexually mature adult stage

midden: the pile of cone scales found on the territories of tree squirrels, usually under a favourite tree

moult: when an animal sheds old feathers, fur or skin in order to replace them with new growth

montane: the ecological zone located below the subalpine; montane regions generally have cooler temperatures than adjacent lowland regions

myccorhizal fungi/myccorhizae: fungi that have a mutually beneficial relationship with the roots of some seed plants

neotropical: the biogeographical region that includes Mexico, Central and South America and the West Indies

neritic: that part of the ocean extending from the low-tide mark to the continental shelf (i.e., to a depth of about 200 m)

nocturnal: active primarily at night

node: a slightly enlarged section of a stem where leaves or branches originate

notochord: a primitive backbone

nutlet: a small, hard, single-seeded fruit that remains closed

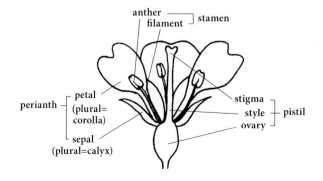

oblanceolate: having a rounded tip that is wider than the base (usually in reference to leaves)

omnivorous: feeding on both plants and animals

operculum: a lid or covering, especially the plate that closes the shell of a gastropod (e.g., a snail) when the animal is retracted

opposite: an arrangement of leaves along a stem in which the leaves are paired and directly opposite one another

ovoid: egg-shaped

palmate: having leaflets, lobes or veins arranged around a single point, like the fingers on a hand (e.g., a maple leaf)

panicle: a branched flower cluster in which the lower blossoms develop first

pappus: the tuft of hairs on a seed (e.g., dandelion or thistle) that aids in wind dispersal

parasitic: a relationship between two species in which one benefits at the expense of the other

patagium: the skin that forms a flight membrane (e.g., in bats and flying squirrels)

pelage: the fur or hair of a mammal

pelagic: open-ocean habitat far from land

perennial: a plant that lives for several years

petal: a member of the inside ring of modified flower leaves, usually brightly coloured or white

petiole: a leaf stalk

photosynthesis: the conversion of carbon dioxide and water into sugars via the energy of the sun

pinniped: a carnivorous, aquatic mammal with a streamlined body specialized for swimming and limbs modified into flippers (e.g., seals, sea lions, walruses)

pinnate: having branches, lobes, leaflets or veins arranged on both sides of a central stalk or vein (e.g., ferns); feather-like

pioneer species: a plant species that is capable of colonizing an otherwise unvegetated area; one of the first species (plant or animal) to take hold in a disturbed area

piscivorous: feeding on fish

pishing: a noise made to attract birds

pistil: the female organ of a flower, usually consisting of an ovary, style and stigma

plastic species: a species that can adapt to a wide range of conditions

plastron: the lower part of a turtle or tortoise shell

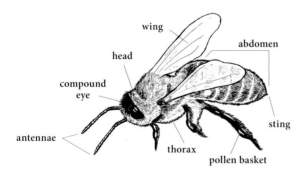

poikilothermic: having a body temperature that is the same as that of the external environment and varying with it

pollen: the tiny grains produced in the anthers of a plant and which contain the male reproductive cells

polyandry: a mating strategy in which one female mates with several males

pome: a fruit with a core (e.g., apple)

precocial: animals that are active at birth or hatching

prehensile: able to grasp

proboscis: the elongated tubular and flexible mouthpart of many insects

raceme: a simple, unbranched inflorescence with flowers on short stalks along an elongated central stalk

ray flower: in a composite flower (e.g., aster, daisy or sunflower), the strap-like outer florets that resemble petals

redd: a spawning nest for fish

rhizome: a horizontal underground stem

rictal bristles: hair-like feathers found around the mouths of some birds

riparian: on the bank of a river or other watercourse

rookery: a colony of nests

rorqual: a type of baleen whale with pleated, expandable skin on its throat

ruminant: a even-toed mammal with a segmented stomach (e.g., cow, moose, deer)

runner: a slender stolon or prostrate stem that roots at the nodes or the tip

rut: the mating season for members of the deer and sheep families, when the stags or rams compete for does and ewes, often engaging in dramatic clashes of antlers or horns

samara: a dry, winged fruit, usually of a maple or ash, with typically only a single seed; also called a **key**

salmonid: a member of the Salmonidae family of fishes, which includes trout, char, salmon and whitefish

saprophyte: a plant or other organism that lives on dead or decaying material

schizocarp: a dry fruit that splits into 2 or more parts at maturity, each part with a single seed

scute: a bony or horny plate (e.g., on a turtle's shell or the underside of a snake)

sepal: the outer, usually green, leaf-like structures that protect the flowerbud and are located at the base of an open flower

sessile: without a stem or stalk

silicle: a fruit of the mustard family (Brassicaceae) that is two-celled and usually short, wide and often flat

silique: a long, thin fruit with many seeds; characteristic of some members of the mustard family (Brassicaceae)

sorus (*pl.* sori): a collection of spore-producing structures on the underside of a fern frond

spadix: a fleshy spike made up of many small flowers

spathe: the leaf-like sheath that surrounds a spadix

spur: a pointed projection

spyhop: a whale behaviour in which the animal, while in an almost vertical position, raises its head out of the water just far enough to look around

stamen: the pollen-bearing organ of a flower

stigma: in a flower, a receptive tip that receives pollen for fertilization

stolon: a long branch or stem that runs along the ground and often propagates more plants

string bog: a bog with raised ridges of vegetation perpendicular to the flow of water and islands of woody plants alternating with sedge mats; also known as a **strong mire**

Subarctic: the region in the Northern Hemisphere that is immediately south of the true Arctic, generally between 50° and 70° N latitude, and covering much of Canada, Alaska, Scandinavia and Siberia

subtend: to be directly below; also to enclose or surround

suckering: a method of tree and shrub reproduction in which shoots arise from an underground runner or spreading roots

syrinx: a bird's vocal organ

taproot: the main, large root of a plant from which smaller roots arise (e.g., a carrot)

tendril: a slender, clasping or twining outgrowth from a stem or leaf

tepal: a collective term used for sepals and petals when there is no clear distinction between the two

terrestrial: land frequenting

territory: a defended area within an animal's home range

test: the external skeleton ("shell") of a sea urchin

torpor: a state of physical inactivity

tragus: a prominent structure of the outer ear of a bat

trifoliate: divided into 3 leaflets

tubular flower: a type of flower with all or some of the petals fused at the base

tundra: a high-latitude ecological zone at the northernmost limit of plant growth, where plants are reduced to shrubby or mat-like forms

tympanum: eardrum; the hearing organ of a frog

umbel: a round or flat-topped flower cluster in which several flower stalks originate from the same point, like the ribs of an inverted umbrella

ungulate: a hoofed animal

utricle: a dry, thin-walled, single-seeded fruit

ventral: of or on the abdomen (belly)

vermiculations: wavy-patterned markings

vertebrate: an animal with a backbone

vibrissae (*sing.* vibrissa): specialized hairs ("whiskers") around the mouths of certain mammals that serve as tactile organs; bristle-like feathers around the beaks of some birds that aid in catching insects

whorl: a circle of leaves or flowers around a stem

woolly: bearing long or matted hairs

wrack line: the line of seaweed, grasses and other debris left on the upper beach by a high tide

INDEX

Names and page numbers in **boldface** type refer to primary species.

ABOUT THE AUTHORS

Erin McCloskey spent her formative years observing nature from atop her horse while growing up in rural Alberta. Erin received her BSc with distinction in environmental and conservation sciences, majoring in conservation biology and management, from the University of Alberta. An active campaigner for the protection of endangered species and spaces, Erin has collaborated with various NGOs and is involved in numerous endangered species conservation projects around the world. Since 2000, she has freelanced as a writer and editor for several magazine and book publishers focused on nature, travel, scientific research and even alternative health care. Erin is the author of *The Bradt Travel Guide to Argentina, Ireland Flying High, Canada Flying High* and *Hawaii from the Air,* and co-author/editor for the *Green Volunteers* guidebook series. Erin is also the author of the *Washington and Oregon Nature Guide, Southern California Nature Guide, Northern California Nature Guide, British Columbia Nature Guide, Bear Attacks* and *Wolves of Canada* for Lone Pine Publishing.

Gregory Kennedy has been an active naturalist since he was very young. He is the author of many books on natural history, and has also produced film and television shows on environmental issues and indigenous concerns in Southeast Asia, New Guinea, South and Central America, the High Arctic and elsewhere. He has also been involved in numerous research projects around the world ranging from studies in the upper canopy of tropical and temperate rainforests to deepwater marine investigations.